# *Urban Environments – History, Biodiversity & Culture*

Edited by Ian D. Rotherham

*International Urban Ecology Review*
Volume 5 2015
Revised Edition 2016
ISSN 1367-7519

ISBN 978-1-904098-62-1

ISSN 1367-7519
ISBN 978-1-904098-62-1

Typeset and processed by Chris Percy & Christine Handley

Published by: Wildtrack Publishing, Venture House, 103 Arundel Street, Sheffield, S1 2NT, South Yorkshire, UK

UKEconet: www.ukeconet.org
Wildtrack Publishing: http://www.ukeconet.org/wildtrack-publishing.html

Front cover picture: Oxeye daisy © Ian D. Rotherham

# Contents

## Editorial Paper

Ian D. Rotherham      Times they are a changin' – Recombinant Ecology as an emerging   1
paradigm

## Main Papers

Mark R. D. Seaward      Triumph over adversity: a new era for urban lichens      20

Ian D. Rotherham      Relict communities and urban commons – urban distinctiveness,   30
history and sustainable urban biodiversity

Annie Chipchase and      Urban ecology and conservation: neglected and undervalued      40
Peter Glaves      subjects?

Melvyn Jones      Urban history and the evolution of townscapes      51

Ian C. Trueman      Issues in urban nature conservation (plants and vegetation)      60

Ian D. Rotherham &      Studying wildlife distribution using 'citizen' science public      68
Mark Walker      sightings confirms how suburban deer are now found throughout
the United Kingdom

David Goode      Some early issues in urban nature conservation      80

## Notes

Mark Bowden      Wastes and strays: the archaeology of urban commons      92

Rachel Remnant      Nature in your neighbourhood: helping local communities      94
improve green spaces for wildlife

Hilary Thomas      Recombinant communities along an urban routeway      96

James Calow      A methodology for enabling companies to establish and      97
implement biodiversity action plans within environmental
management systems

Shona Turnbull      Biodiversity in the urban community: the importance of      98
involving local people in wildlife projects

Peter Shaw      Nature reserves, orchids and fly ash      102

## Book Reviews      104

*IUER Volume 5*

# *Urban Environments – History, Biodiversity & Culture*

## Foreword

This volume is a retrospective publication of contributions originally to two national conferences / seminars held in Sheffield, on the theme of 'Urban Environments – History, Biodiversity and Culture'. To the updated papers from those events we have added invited current contributions on the themes of urban nature and urban ecology. Ideas and issues in urban ecology become more significant as globalisation, urbanisation and cultural severance shape our world and our future ecologies. This is paralleled by increasing interest in the underpinning science and research paradigms in relation to urban environmental spaces. In the early 2000s, the short-lived Research Council-funded 'URGENT' programme was largely a case of paying to tell us what most people working in urban ecology already knew. Ecologists new to the urban context suddenly became excited about the juxta-position of pollution and biodiversity in degraded and contaminated sites, something well-known to urban ecologists and naturalists since the 1980s or earlier. Similarly, the contributions of urban gardens to nature conservation were greeted with surprise and excitement. However, what was missing then, and still largely absent in contemporary urban ecological studies, is the human historical dimension. This was something brought to the fore so eloquently by the late Oliver Gilbert.

With the recent publication of David Goode's New Naturalist volume 'Nature in Towns and Cities', there has been a resurgence in the excitement about urban nature. With the on-going urbanisation of the planet, these issues are set to grow rather than to diminish, and the emerging concepts of 'recombinant ecology', 'ecological fusion', and 'ecological hybridisation', provide conceptual frameworks for future investigation.

Ian D. Rotherham, October 2015

# *Times they are a changin' – Recombinant Ecology as an emerging paradigm*

**Ian D. Rotherham,**
Sheffield Hallam University

*Editorial Paper*

## Abstract

Invasive and exotic species have long been of interest to ecologists and natural historians (for example Salisbury, 1961; Fitter, 1945; and Lever, 1977). However, the ecological effects, beyond the immediacy of invasion and species displacement are rarely considered, though with the work of Gilbert (1989, 1992a, 1992b), and of Barker (2000), this began to change. Across the globe, communities are becoming urbanised and the rate of urbanisation is growing. Since the early 2000s, over half the world's human population has been urban. This triggers issues of cultural severance in rural areas which remain, but also other issues for ecology and ecosystems as they respond to change. With global transport and communications there is a '*Disneyfication*' of ecology as species are transported across and around the planet. However, it is mostly in and around the cities and towns that the processes of introduction, of mixing and of invasion take precedence. This ecological mixing I refer to as '*eco-fusion*' and it generates new hybrid ecologies. However, it should not be assumed that these processes are either new, or essentially urban. Recombination of ecology has been an on-going phenomenon ever since people themselves colonised and manipulated the landscape. Only now, in the age of the Anthropocene (Steffen *et al.*, 2007), the scale of human domination over nature and with the advent of a truly global, urban society has the process come to the fore. This paper takes a largely British case study approach to this important, emerging paradigm in ecology.

## Introduction

Hybridisation of nature has been driven by long-term nature-human interactions. These occur in agriculture and forestry, and increasingly occur through urbanisation and related environmental change (Freedman, 2005). Changing ecology and ecosystems are a result of urbanisation, globalisation, climate change, and human cultural influences and the consequences are subject to ongoing research and debate (e.g. Johnson, 2010; Hobbs *et al.*, 2013; Prins & Gordon, 2014, and many others). Accelerating globalisation, and both human-induced and natural climate change, speed the hybridisation process. Human impacts include disturbance, nutrient enrichment, habitat replacement (formation and destruction), and planetary-scale species dispersal (Rotherham, 2014a; Douglas *et al.*, 2011). However, the ecology driving these changes is the 'natural' mechanism of ecological succession and change, with consequent hybridisation and adaptation of species and ecosystems.

A key factor today is that species mixing is at a rate unprecedented in the history of biodiversity evolution and is generating obviously novel ecologies (Hobbs *et al.*, 2013; Jørgensen *et al.*, 2013). Now recognised as the '*Anthropocene*' (Steffen *et al.*, 2007), in a new evolutionary epoch nature is adapting to this new canvas and altered template. A major result is the advent of ecological fusion or '*eco-fusion*' as a dynamic, ongoing process of interactions as species '*native*' or '*alien*' to particular locations or regions form newly combined ecological communities. Species may enter these novel communities whilst others, are displaced (Jørgensen *et al.*, 2013; Hobbs *et al.*, 2013). The 'Anthropocene', is typified by human-driven influences to which nature responds (Steffen *et al.*, 2007; Rotherham, 2014a). Given this scenario, it is increasingly important to understand environmental history to provide context to mainstream ecology. This helps to further understanding of the drivers of these changes and to improve the predictability of future ecological outcomes (Hall, 2009; Jørgensen *et al.*, 2013; Rotherham, 2014; Samojlik *et al.*, 2013; Smout, 2000). The need to understand better, the evolving ecologies and thereby inform planning processes has been growing over recent decades (e.g. Douglas *et al.*, 2011; Forman, 2014; Hough, 1995; Sukopp *et al.*, 1995).

The context of these changes is one of critical issues of ecology and ecosystems under stress at every level from local, parochial, to global and planetary (Adams, 2003; Barker *et al.*, 1994; Gaston, 2010; Rotherham, 2014a). These processes are driven by globalisation, climate change, urbanisation, and other human cultural influences (Niemelä, 2011; Sukopp & Hejny, 1990; Agnoletti, 2006; Agnoletti *et al.*, 2007). Importantly, many changes are predictable through the application of informed knowledge of ecosystems and of species strategies (e.g. Grime *et al.*, 2007; Hodgson, 1986; Rotherham, 2014).

Ecological diversity reflects underlying biological and ecological processes around the world. For a particular locale, for example Britain, diversity results from matrices of geographical spaces (habitats with parochial environmental conditions with diverse stages and states of flux and stability), and total '*biodiversity*' is a summation of this. With innumerable sites, both species-rich, species-poor, this forms the national ecology. Importantly though, it is not fixed, but a shifting, drifting, fluxing, human-influenced, nature-influenced, climate-influenced, resource. In a context of time and space, this resource is not a fixed or finite entity. Furthermore, over immensely long timescales, evolutionary processes generate new species and drive others to extinction. Geological forces cause massive movements of continents with extinctions and associated phases of rapid evolution. In lesser timescales, periods of glaciations, inter-glacials and ice-ages, stress and test ecological systems. All the other changes happen against this planetary backdrop, a '*broader canvas*' of dynamic, shifting ecologies (Rotherham, 2014b).

**Figure 1. The general process of eco-fusion [Adapted from Rotherham, 2016a]**

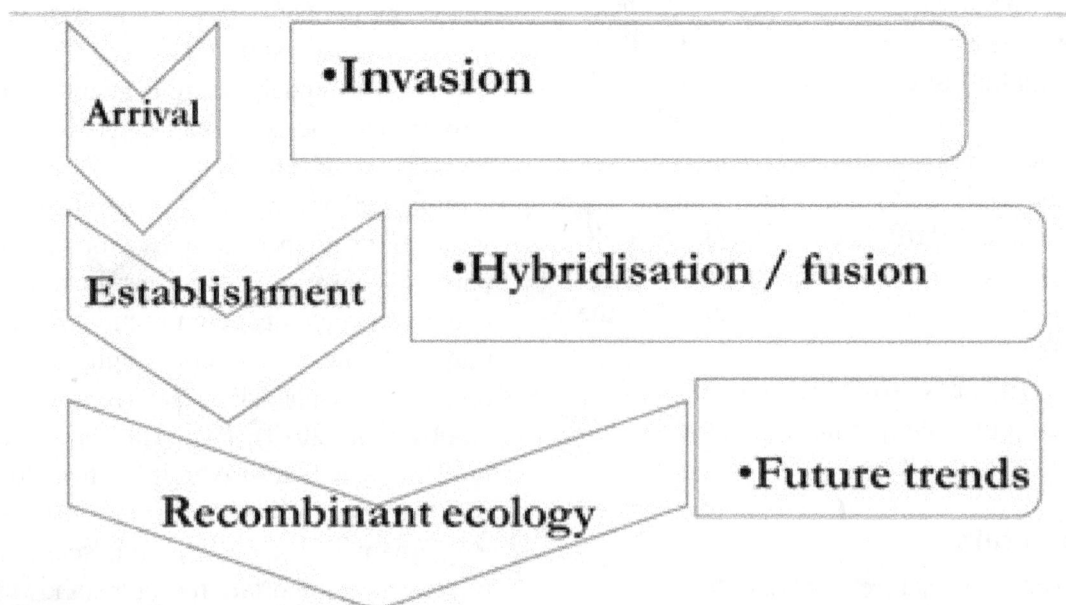

# A cultural facilitation of invasion

Throughout history people have interacted with nature to modify and sometimes destroyed environmental resources (Rackham, 1986; Rotherham, 2013a; Rotherham, 2014a). Human management has generated identifiable and distinctive '*cultural*' landscapes, as fusions of natural and anthropogenic elements. Furthermore, many of these lands have been managed with customary and traditional mechanisms. In a pre-petrochemical age this management created traditional landscapes frequently made up of diverse, species-rich habitats and maintained by long-established land-use applied continually over the years. It is argued (Rotherham, 2014a, 2016), that such traditional, 'unimproved' lands provided habitats for biodiversity with direct links to habitat analogues in the ancient, primeval, '*natural*' European landscape (Rotherham, 2009b, 2013b, 2014a), the vision of Frans Vera (Vera, 2000). I suggest that the complex inter-relationships and ecologies of, for example, species-rich limestone grasslands, fens, peatlands, and ancient woods, result from such long-lived human-nature interactions and they are analogues of the ecological communities of the primeval landscape.

The human influences in these ecosystems include the hybridisation of species and ecology, '*eco-fusion*'. This is most readily recognised in the world's increasingly urban environments, but is clearly occurs more widely. Furthermore, whilst recombination as a process has only been recognised in recent decades, it has occurred on varying scales since humans first impacted on the landscape. Today, huge areas of land are dominated by forestry and agriculture and in these imposed environments plants, animals, and fungi move and mix beyond natural distributions and limits. This means that old, new, native, and exotic, are intertwined in novel, recombinant communities in hybrid ecosystems. Particularly now, in the rapidly expanding urban heartlands, this new ecology, of native

**Figure 2. The flow of recombinant ecology [Adapted from Rotherham, 2016a]**

## But the processes are not new – we can see them in landscape-scale changes over centuries

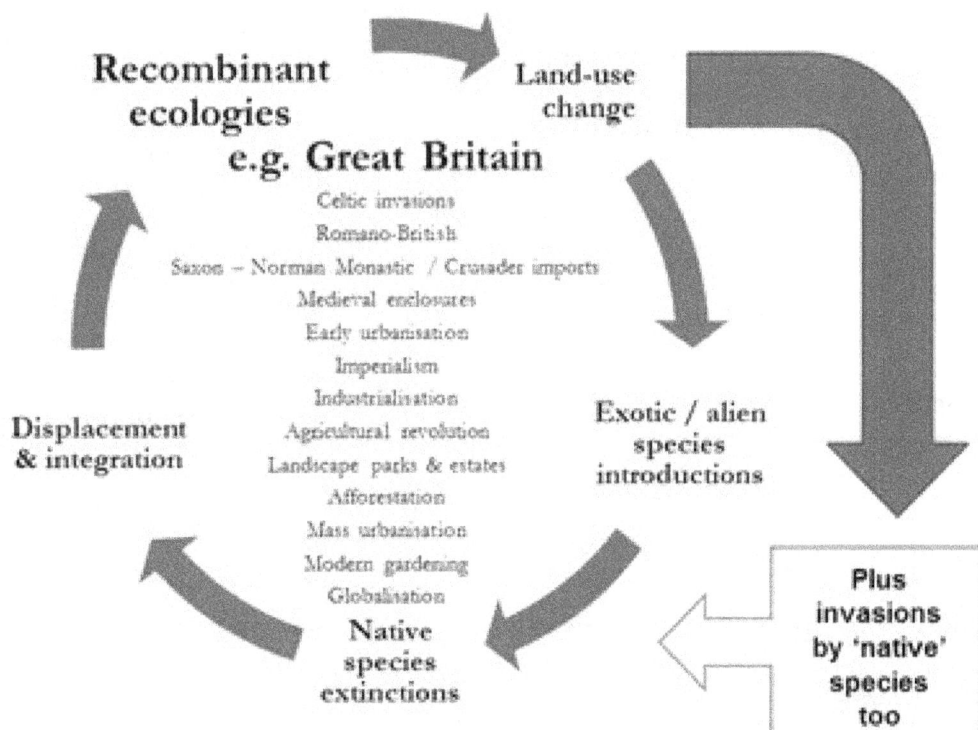

and alien, is locked in a perpetual and dynamic struggle for dominance with the resulting formation of novel dependencies, interactions, and communities (Rotherham, 2014a, 2016; Gilbert, 1989, 1992a). Recognition of these processes and their consequences challenges many current debates in conservation ecology, particularly in relation to debates on alien species. The approach generates new paradigms and matters of perceptions, judgements, and actions. These ideas of recombinant, hybrid ecologies and eco-fusion processes are new, and they have significant implications for future ecologies. Indeed, for discussions on ideas of '*wilding*' and '*re-wilding*', ecological fusion and hybrid ecology, are vital conceptual frameworks for debates on future landscapes and ecologies (Rotherham, 2014a, 2014b, 2016; Prins & Gordon, 2014; Taylor, 2005).

## Hybrid Ecologies

Over centuries, non-native, introduced species have altered landscapes and indigenous ecologies (Gilbert, 1989; Rackham, 1986; Rotherham & Lambert, 2011; Rotherham, 2014a, 2014b). Furthermore, these changes, such as by the rabbit introduced by the Normans to Britain, may be the fundamental determinants and keystone species in modern-day ecology. However, trends and changes need to be assessed in the wider context of fluxing climate, land-use, and other human-nature interactions. A brief consideration of British landscapes shows there is little genuinely pure '*natural*' or '*native*'. Most of our environment is dominated by '*cultural*' or '*eco-cultural*' landscapes (Rotherham, 2007a, 2008a, 2011, 2009b, 2014a). This ecology results from centuries of interactions between people and nature, with '*semi-natural*' components and some traditional landscapes. Large areas of British farmland for example, are of modern origins and ecologically highly exotic in their make-up.

Whilst natural changes are influential, people are at the core, directly or indirectly, of most biological invasions (Johnson, 2010; Rotherham & Lambert, 2011). Two particular

examples of invasion and fusion can be witnessed in Great Britain to confirm the human component of the human-nature influence. The first example is the deliberate introduction of plants and animals around the recently discovered world, by the Victorian Acclimatisation Societies. The second example is the often-overlooked, case of the Victorian Wild Garden Movement (Rotherham, 2005, 2014a, 2016; Rotherham & Lambert, 2011). In these two nineteenth-century phenomena we see the germ of many of contemporary issues and challenges for modern nature conservation and land management. There is also an interesting twist in terms of the changing perceptions, attitudes, and politics in relation to nature and the exotic. This is often overlooked but is hugely influential from the early nineteenth century to the early twenty-first century. Davis *et al.* (2001) wrote of changing British attitudes to exotic species with the seminal writings and broadcasts of Charles Elton. However, the more general and pervading influences of fashion and taste with regard to exotic species has been ignored. Indeed, the often critical roles of accidental or even deliberate introductions of now invasive species have been overlooked (See Rotherham, 2001, 2005a, 2005b) or not recognised.

Human cultural facilitation of invasions is very significant. Furthermore, perceptions and attitudes have influenced the processes and the responses (e.g. Rotherham & Lambert, 2011). However, the research in this field is often lacking since the work inherently crosses the boundaries of ecological science and of history. Furthermore, it is obvious that many if not most of the ecological invasions of the twentieth, and twenty-first centuries in Britain were culturally facilitated (e.g. Rotherham, 2001, 2005a, 2005b, 2009a; Rotherham & Lambert, 2011). Indeed, the problems associated with aggressive and invasive plants and animals are not new (for example the spread of both black rats and brown rats, and their impacts on people and landscapes), but the scale of impact, now combined with rapid climate change and other environmental influences, is spectacular. It is suggested that 15% of Europe's 11,000 exotic species combine to generate environmental and economic damage valued at £2bn per annum to the UK economy. However, behind the often-torrid newspaper headlines remain hey questions about what is native and where, what is alien, and when; from Spanish bluebells, to eagle owls, Canada geese, ruddy ducks, ring-necked parakeets, Japanese knotweed, Himalayan balsam, to feral big cats, beavers and signal crayfish, and wild boar. Which ones get a '*free pass*' to a new hybrid future and which do not (Rotherham, 2009a, 2014b; 2016; Rotherham & Lambert, 2011)?

## Global empires, acclimatisation & wild gardening

Until the 1940s and the aftermath of the Second World War, Britons travelled the world seeking out and collecting new species to take home. Indeed, they went a step further and implemented deliberate programmes of release into the countryside to '*improve it*'. This was intended to bring about economic gain on the one hand and to beautify the landscape on the other. The ideas spread and a movement emerged and was formalised. This fashion then spread around the world, firstly with European Acclimatisation Societies, aiming to introduce and test new economic crops and to investigate the potential for food. In Britain and the British colonies, the Acclimatisation Societies sought to introduce animals and birds to new territories so as to better their economies, gastronomies, and landscapes (Lever, 1977; Rotherham, 2011b). These Victorian Acclimatisation Societies had massive impacts around the planet in places such as New Zealand, Australia and the West Indies for example. Indeed, in many places the introduced species transformed the landscape and eliminated many natives.

In Britain in particular, with an imperial domination of the globe and at home a growing penchant for gardening and landscaping, the scene was set for a radical transformation of the countryside and its ecology. The Victorian 'Wild Gardeners' spread plants like rhododendron (introduced in the previous century), Himalayan balsam, Japanese knotweed, and giant hogweed, widely across the British countryside. They were aided and abetted by foresters and estates managers who rushed to adopt exotic species, and in forestry, the European fashion for high forest plantations. Garden designer and writer William Robinson formalised and popularised the idea of '*naturalising*' exotic plants into landscapes instead of merely planting them for effect (Robinson, 1870). Undoubtedly the use of exotics was already widespread, but Robinson made it mainstream and popular in the 1800s and early 1900s.

Some of these species were recognised locally as invaders at a relatively early date, for example Himalayan balsam in Manchester by the mid-1800s (e.g. Grindon, 1859), but only as conservation problems much later, by the 1970s and 1980s. However, they have now been joined by a whole raft of others such as *Montbretia*, *Buddleia*, and Cherry Laurel. It seems that these escapees spread across the land and build slowly over perhaps 50-100 years as they colonise suitable environments. However, they then move into an exponential and explosive expansion at which point botanists and others begin to notice and record them. By this time it is of course almost impossible to control further spread.

More recently recognised as an invasive species is the variegated yellow archangel which I now believe to be a Victorian cultivar, which is spreading rapidly whenever introduced to woodland or hedgerow. It can cover several hundred square metres of woodland ground flora in only 10-15 years, and since it does not produce viable seed, such spread is totally vegetative, and its arrival is always by direct human agency. It seems that the role of '*Wild Gardeners*' has not been recognised in triggering twentieth-century plant invasions, yet with the exception of the massive establishment of exotic conifers for forestry, it lies at the heart of most terrestrial plant naturalisations in Britain. This was first noted by Rotherham (2001, 2005a, 2005b), but the reader should seek out

William Robinson's hugely influential book *The Wild Garden* (1876). The most influential popular writer on English gardening in the late 1800s, this landmark volume sparked a revolution in garden taste over the subsequent fifty years or so, and associated with that, an almost unprecedented episode of ecological recombination. Allan (1982) documented and described Robinson's life and work.

From the 1700s to the early 1900s, releases of alien plants were firstly into domestic and landscape gardens, but also into forest and woodland estates. It seems that inspired by William Robinson, the Victorian Wild Garden movement was responsible for the introduction and subsequent escape of many plants into the British countryside. Indeed, these include some of the most spectacular invaders such as giant hogweed, and giant knotweed; and these are truly stunning garden plants for the 'Wild Garden' of the late 1800s. In the mid-Victorian period, when William Robinson published his book '*The Wild Garden*' in 1870, across the British countryside and in suburban gardens practices he described and advocated were already well established. In particular, these included the embellishment of woodlands, utilising naturalisation of exotics in landscapes, and mixing plants from sub-tropical gardens with emphasis on strong foliage groupings; the impacts remain with us today. However, with the advocacy of Robinson, this became basis of '*wild gardening*' during the 1870s and 1880s (Elliott, 1986). The approach involved large scale, establishment of species capable of spreading themselves in large masses. In this way, plants like *Rhododendron ponticum*, *Heracleum mantegazzianum* (giant hogweed), *Polygonum* sp. (Japanese and giant knotweeds), and Himalayan balsam, were all excellent choices. However, these same wild garden favourites are now the scourge of nature conservation. In situations like Chatsworth Park in Derbyshire or Clumber Park in Nottinghamshire, landscape designers juxtaposed exotic garden plants with forest trees and wild undergrowth. Today at Chatsworth for example, the rhododendrons, balsam and even giant hogweed remain. William Robinson asserted that the principle

of wild gardening was '... *naturalizing or making wild innumerable beautiful natives of many regions of the earth in our woods, wild and semi-wild places, rougher parts of pleasure grounds, etc.*' Today, an assessment of British exotic flora confirms how many of today's most problematic alien '*weeds*' have their origins in deliberate introductions. With globalisation, urbanisation and climate change, the escape of exotic garden plants continues apace.

Significantly, in the wider landscape during the same period, there was a change from traditional coppice management of native woods to high forestry. Commons were enclosed and commoners displaced, and a new, intensive, modern farming landscape was imposed over large areas. In woodlands, this new approach was often with exotic tree species, supplemented by introduced shrubs such as rhododendron, snowberry, *Mahonia*, and *Gaultheria*, and other species brought from around the world. Many of these were introduced for woodland management linked to game preservation. Indeed, this period of the 1700s to the early 1900s was perhaps a golden age of plant introductions to the British Isles; there is no going back. Since then there has been a continual release of new exotic species into Britain's landscape from gardens and from modern plantings of exotic trees and shrubs. Indeed, from the latter, a wide range of berry-bearing trees and shrubs is rapidly establishing throughout woods and forests. Ratcliffe (1984) noted the beginnings the spread of shrubs like *Berberis* and *Cotoneaster*; and exotic cultivars of holly and *Sorbus* are now spreading widely into woodlands.

## A history of plant importation

The *Acclimatisation Societies* and the *Victorian Wild Gardeners* (Rotherham, 2005a, 2005b) were manifestations of processes that evolved in varying degrees over centuries. Throughout history waves of settlers or conquerors of the British Isles had brought new plants and animals with them. The Celts, the Romans, and then Normans and later the Crusaders, for example, imported animals and plants, many of which became keystone species of modern ecologies. Most obvious is the humble rabbit which impacts hugely on modern landscapes. However, fallow deer and brown hare are others and these raise interesting issues of perception and attitude (Rackham, 1986). Alongside deliberate introductions, were many other animals and plants which escaped domestication to make their own way in the globalised world. Many such species have become intimate components of what is now considered '*British*' ecology. Romans and Normans brought herbs and food-plants from southern Europe and the Mediterranean, as did Crusaders and various monastic dynasties which controlled large areas of Britain's productive countryside over several centuries. Over this period, many species were absorbed or hybridised into native ecology. Interestingly, most of these species are now tolerated, and many (like the brown hare for example), are celebrated and conserved as 'honorary natives'.

The process continued unabated throughout the later medieval period since by the 1500s, traders and seafarers from Britain and Holland in particular, were charting and colonising the world. From their visits they brought back exotic plants and sometimes animals; many introductions perishing but not, and accidental imports already included black rat and brown rat, with the added bonus of bubonic plague. In return, later explorers spread dogs, cats, and much more; doing untold damage to previously isolated island ecologies. The cultural homogenisation of ecology was accelerating with the collection and dissemination of exotic species as travellers sought new plants and animals for gardens and menageries. Then, as landscaping, forestry, and gardening erupted in Britain during the 1700s and 1800s, environmental impacts increased; a process continuing today sometimes with catastrophic results (Rotherham, 2014a).

## Landscapes transformed

These changes to 'native' ecology were not isolated from other impacts. Indeed, at about this time, the eighteenth and nineteenth centuries, wider landscapes in Britain were traumatised by parliamentary enclosures. Commonland was wrested from commoners, peasants and the poor, and converted into intensive food production units; a process continuing to the present day. Natural or traditional countryside and ecology were swept away by the tide of change (Rotherham, 2014a). For example, traditional coppice woods were converted to high forest plantations, and industrialising cities sprawled over the countryside (Rackham, 1980, 1986; Rotherham, 2014a). In the areas which remained relatively intact, increasingly populated by plants and animals from around the world, lands that remained morphed into leisurely landscapes for the pleasure of landowning industrialists (Rotherham, 2014a).

A result of these changes was the transformation from ecology dominated by native '*stress tolerators*', to exotic species which were mostly '*ruderal*' and '*competitive*' plants. The trend was noted for the twentieth century by Davis *et al.* (2001), but it really began much earlier with landscapes flexing and changing. Disturbance and nutrient enrichment (eutrophication) emerged as dominant influences (Rotherham, 2014a).

## Transformed ecologies

An uncomfortable truth (Rotherham, 2016), based on ecological history and compounding effects of the cessation of traditional and customary countryside practices is a radically transformed ecology. This process increased in significance throughout the 1800s into the late 1900s. The backdrop to this is '*cultural severance*' (Rotherham, 2009b; 2013), which equates to the ending of traditional and customary uses, values and management of (mostly) rural and ecological resources. For many landscapes, release from subsistence exploitation of centuries has meant rapid increase in biomass and nutrients, whilst stress tolerant species (often of high

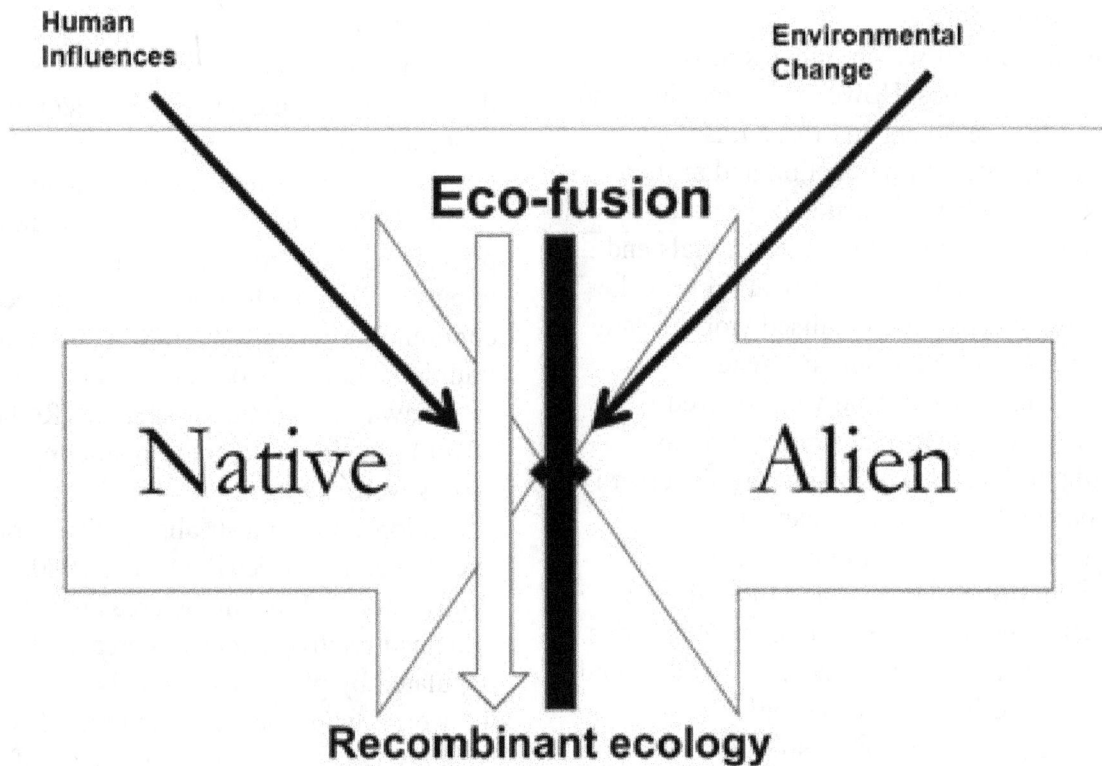

**Figure 3. Influences on eco-fusion of natives and aliens to produce recombinant ecologies [Adapted from Rotherham, 2016a]**

conservation value) go into rapid decline (Webb, 1986, 1998). The result of modern-day changes is either abandonment or pulses of macro-disturbance. These replace the micro-disturbance associated with traditional management (Rotherham, 2009a, 2014a).

## Recombinant ecologies

It is clear that these ecosystem stresses are most obvious in urbanised zones. Here, combined with exotic species as described earlier, these species form new ecological associations, a '*recombinant ecology*' (Barker, 2000; Rotherham, 2014a). The emerging communities are different and distinctive from what went before and long-term trends in vegetation for example, can be recognised at regional scales (e.g. Hodgson, 1986). Work by for example Hodgson (1986), and Grime *et al.* (2007) confirmed the declines and replacements expected from cultural severance.

There are elements of former landscapes in '*semi-natural*' (or eco-cultural) habitats. However, even here cultural drivers over

centuries of human exploitation have changed and now often ended (Rotherham, 2009b, 2014a). In some case there is subtle, long-term blurring of ecology, but in others, changes are rapid and dramatic. Considering the changes it is remarkable that so much former landscape and ecology are visible in the modern veneer. Indeed, some elements of the ancient ecosystems are surprisingly resilient. That is unless they are totally swept aside by modern mechanisation (Rotherham *et al.*, 2013). However, despite this, there are major issues for conservation. In Britain this includes the latter-day recognition of medieval parks for example, as important remnants of the '*Frans Vera primeval landscape*' (Rotherham 2007b; Vera 2000). For decades, these areas received little recognition or protection.

In current debates on wilding (or so-called, re-wilding), decides to abandon sites to feral nature or intervene with planned release of large herbivores raise many issues. In order to allow trees freedom to regenerate, do we fence out of wild (feral) herbivores such as the Scottish Highland red deer. These are all human-determined interventions (Rotherham, 2014b) and Ayres (2013), for example, welcomed that '*When you let go of control of the land and let nature run its course it is unpredictable, often with surprising and positive outcomes.*' This is fine in principle, but what happens if you get a long-term, dominant bracken stand rather than the bluebell woodland that is promised? Should we control feral red deer numbers or let nature take its course of animal starvation and depauperate woods without tree regeneration? What happens when the last remnants of rich biodiversity are lost? If we are to intervene, then who does it, why do they do it, what do they do, where do they do it, and when do they do it? Furthermore, of course, who decides and who pays? Will land managers, conservationists, and even the public, accept exotic plants such as rhododendron, sycamore, larch, spruce, Japanese knotweed, Himalayan balsam and giant hogweed spreading feral recombinant ecologies across the landscape. Free-willed, recombinant nature mixes these species with mink, rabbit, grey squirrel, Canada goose, ruddy duck, ring-necked parakeet, signal crayfish, and exotic or feral deer (Rotherham, 2014b). Recombinant ecology emerges from ecological fusion processes, but as yet, many conservationists remain reticent to accept such changes. It is highly likely that feral ecologies will be dominated by this heady recombinant mix of exotic and native invasive (like birch and bracken).

## Science, politics & environmental democracy

Issues of alien invasive and exotic species need to be considered within the wider setting of environmental change, conservation, and politics (Rotherham & Lambert, 2011; Rotherham, 2016). We already have hybrid ecology and this situation will inevitably grow. In this context, ideas of recombination and eco-fusion are increasingly significant (Rotherham, 2014a).

## Conclusions: emerging paradigms & new concepts

It is clear that ideas of ecological stability depend on time-scale but also on presumptions of long-term environmental stasis. These assumed truths are wrong and even considering short periods, environmental conditions and therefore ecologies, fluctuate, sometimes markedly. Furthermore, human transformation of environmental conditions has changed ecology in the past and continues to do so today but at an increasing scale and pace. Combined with the changed environmental baselines, people also move species of animals and plants around the landscape and across the world. This is constantly creating new potential mixes of ecologies with ecological fusion professes generating novel, recombinant communities. With globalisation and urbanisation increasing dramatically, these observations have huge significance for nature conservation.

In particular, discussions of new ways to manage nature and the landscape need to engage with the emerging paradigms of environmental and ecological history. Long-term studies of ecological history in Britain, for example, confirm the hybrid nature of our '*native*' ecology. They also indicate the increasingly hybrid character of our future nature. This hybridisation, so-called 'eco-fusion', occurs at the level of the community as aliens and natives mix and merge into recombinant ecologies, but also at that of the species as hybrids emerge from both deliberate and accidental fusion. I develop this theme in a later publication – see for example, Rotherham (2016b).

## References

Adams, W. (2003) *Future Nature: a vision for conservation.* Earthscan, London.

Agnoletti, M. (ed) (2006) *The Conservation of Cultural Landscapes.* CAB International, Wallingford, Oxon, UK.

Agnoletti, M., Anderson, S., Johann, E., Kulvik, M., Saratsi, E., Kushlin, A., Mayer, P., Montiel, C., Parrotta, J., & Rotherham, I.D. (2007) *Guidelines for the Implementation of Social and Cultural Values in Sustainable Forest Management: A Scientific Contribution to the Implementation of MCPFE - Vienna Resolution 3*. IUFRO Occasional Paper No. 19, ISSN 1024-414X, IUFRO Headquarters, Vienna, Austria.

Ayres, S. (2013) The feral book – reintroducing rewilding. *ECOS*, **34**(2), 41-49.

Barker, G. (ed.) (2000) *Ecological recombination in urban areas: implications for nature conservation*. English Nature, Peterborough, 21-24.

Barker, G., Luniak, M., Trojan, P., & Zimny, H. (eds) (1994) Proceedings of the Second European Meeting of the International Network for Urban Ecology. *Memorabilia Zoologica,* 49, Warsaw.

Bornkamm, R., Lee, J.A., & Seaward, M.R.D. (eds) (1982) *Urban Ecology*. Blackwell Scientific Publications, Oxford.
Davis, M.A., Thompson, K., & Grime, J.P. (2001) Charles S. Elton and the dissociation of invasion ecology from the rest of ecology. *Diversity and Distribution,* 7, 97-102.

Douglas, I., Goode, D., Houck, M.C., & Wang, R. (eds) (2011) *The Routledge Handbook of Urban Ecology*. Routledge, London & New York.

Elliott, B. (1986) *Victorian Gardens*. B.T. Batsford, London.

Fitter, R. (1945) *London's Natural History*. Collins New Naturalist, London.

Freedman, B. (1995) *Environmental Ecology – The Effects of Pollution, Disturbance and Other Stresses*. Second Edition, Academic Press, San Diego.

Gaston, K.J. (ed.) (2011) *Urban Ecology*. Cambridge University Press, Cambridge.

Gilbert, O.L. (1989) *The Ecology of Urban Habitats*. Chapman and Hall. London.

Gilbert, O.L. (1992a) *The flowering of the cities....The natural flora of 'urban commons'*. English Nature, Peterborough.

Gilbert, O.L. (1992b) *Rooted in stone. The natural flora of urban walls*. English Nature, Peterborough.

Grime, J.P., Hodgson, J.G., & Hunt, R. (2007) *Comparative Plant Ecology. A Functional approach to common British species*. Second Edition. Castlepoint Press, Dalbeattie.

Grindon, L.H. (1859) *Manchester Flora*. [No publisher or location given].

Hobbs, R.J., Higgs, E.S., & Hall, C.M. (eds) (2013) *Novel Ecosystems. Intervening in the New Ecological World Order*. Wiley-Blackwell, Chichester.

Hodgson, J.G. (1986) Commonness and Rarity in Plants with Special Reference to the Sheffield Flora. *Biological Conservation,* **36**(3), 199-252.
Johnson, S. (ed) (2010) *Bioinvaders*. White Horse Press, Cambridge.

Jørgensen, D., Jørgensen, F.A., & Pritchard, S.B. (eds) (2013) *New Natures. Joining Environmental History with science and technology studies*. University of Pittsburgh Press, Pittsburgh.

Lever, C. (1977) *The Naturalized Animals of Britain and Ireland*. Hutchinson & Co (Publishers) Ltd, London.

Monbiot, G. (2013a) *Feral: Searching for enchantment on the frontiers of rewilding*. Allen Lane, London.

Monbiot, G. (2013b) The Lake District is a wildlife desert. Blame Wordsworth. *The Guardian,* Monday 2 September.

Niemelä, J. (ed) (2011) *Urban Ecology*. Oxford University Press, Oxford.

Rackham, O. (1980) *Ancient Woodland: its history, vegetation and uses in England.* Edward Arnold, London.

Rackham, O. (1986) *The History of the Countryside.* Dent, London.

Ratcliffe, D.A. (1984) Post-medieval and recent changes in British vegetation; the culmination of human influence. *New Phytologist,* **98**, 73-100.

Robinson, W. (1870) *The Wild Garden.* The Scolar Press, London.

Rotherham, I.D. (1999) Urban Environmental History: the importance of relict communities in urban biodiversity conservation. *Practical Ecology and Conservation,* **3** (1), 3-22.

Rotherham, I.D. (2001) *Himalayan Balsam - the human touch. In: Bradley, P. (Ed.) Exotic Invasive Species -should we be concerned?* Proceedings of the 11th Conference of the Institute of Ecology and Environmental Management, Birmingham, April 2000. IEEM, Winchester, 41-50.

Rotherham, I.D. (2005a) Invasive plants – ecology, history and perception. *Journal of Practical Ecology and Conservation Special Series,* No. 4, 52-62.

Rotherham, I.D. (2005b) Alien Plants and the Human Touch. *Journal of Practical Ecology and Conservation Special Series*, No. 4, 63-76.

Rotherham, I.D. (2009a) Exotic and Alien Species in a Changing World. *ECOS,* **30** (2), 42-49.

Rotherham, I.D. (2009b) *The Importance of Cultural Severance in Landscape Ecology Research.* In: Dupont, A., & Jacobs, H. (eds) (2009) *Landscape Ecology Research Trends.* Nova Science Publishers Inc., USA.

Rotherham, I.D. (2011a) *A Landscape History Approach to the Assessment of Ancient Woodlands.* In: Wallace, E.B. (ed) (2011) *Woodlands: Ecology, Management and Conservation.* Nova Science Publishers Inc., USA, 161-184.

Rotherham, I.D. (2011b) In: Rotherham, I.D., & Lambert, R.A. (eds) (2011) *Invasive and Introduced Plants and Animals: Human Perceptions, Attitudes and Approaches to Management.* EARTHSCAN, London.

Rotherham ID (2013a) *The Lost Fens: England's Greatest Ecological Disaster.* The History Press, Stroud.

Rotherham, I.D. (ed) (2013b) *Cultural Severance and the Environment: The Ending of Traditional and Customary Practice on Commons and Landscapes Managed in Common.* Springer, Dordrecht.

Rotherham, I.D. (ed) (2013c) *Trees, Forested Landscapes and Grazing Animals: A European Perspective on Woodlands and Grazed Treescapes.* EARTHSCAN, London.

Rotherham, I.D. (2014) *Eco-history: An Introduction to Biodiversity and Conservation.* The White Horse Press, Cambridge.

Rotherham, I.D. (2014) The Call of the Wild. Perceptions, history people & ecology in the emerging paradigms of wilding. *ECOS,* **35**(1), 35-43.

Rotherham, I.D. (2016a) *Eco-fusion of alien and native as a new conceptual framework for historical ecology.* Chapter in edited volume for Springer, edited by Arnaldo Melo, Cristina Joanaz, & Estelita Vaz, Springer, Dordrecht, The Netherlands.

Rotherham, I.D. (2016b) *Recombinant Ecology – a hybrid future?*, Springer Briefs, Springer, Dordrecht, The Netherlands.

Rotherham, I.D., Handley, C., Agnoletti, M., & Samoljik, T. (eds) (2013) *Trees Beyond the Wood – an exploration of concepts of woods, forests and trees.* Wildtrack Publishing, Sheffield.

Rotherham, I.D., & Lambert, R.A. (eds) (2011) *Invasive and Introduced Plants and Animals: Human Perceptions, Attitudes and Approaches to Management.* EARTHSCAN, London.

Rotherham, I.D., Lunn, J., & Spode, F. (2012) Wildlife and Coal – the nature conservation value of post-mining sites in South Yorkshire. In: Rotherham, I.D., & Handley, C. (eds) (2012) *Dynamic Landscape Restoration. Landscape Archaeology and Ecology Special Series.* Papers from the Landscape Conservation Forum, (1), Wildtrack Publishing, Sheffield, 30-64.

Salisbury, E. (1961) *Weeds and Aliens.* Collins, London.

Smout, T.C. (2000) *Nature Contested: Environmental History in Scotland and Northern England Since 1600.* Edinburgh University Press, Edinburgh.

Steffen, W., Crutzen, P.J., & McNeill, J.R. (2007) The Anthropocene: Are Humans Now Overwhelming the Great Forces of Nature. *AMBIO*, **36**(8), 614-621.

Sukopp, H., & Hejny, S. (eds) (1990) *Urban Ecology. Plants and Plant communities in urban environments.* SPB Academic Publishing. bv, The Hague, The Netherlands.

Sukopp, H., Numata, M., & Huber, A. (eds) (1995) *Urban Ecology as the basis of urban planning.* SPB Academic Publishing. bv, The Hague, The Netherlands.

Taylor, P. (2005) *Beyond Conservation. A wildland strategy.* Earthscan, London.

Vera, F. (2000) *Grazing Ecology and Forest History.* CABI Publishing, Oxon, UK.

Webb, N.R. (1986) *Heathlands.* Collins, London.

Webb, N.R. (1998) The traditional management of European heathlands. *Journal of Applied Ecology,* **35**, 987-990.

Ian D. Rotherham is Professor of Environmental Geography and Reader in Tourism & Environmental Change in the Department of the Natural & Built Environment, Sheffield Hallam University. i.d.rotherham@shu.ac.uk; www.ukeconet.org

Above: Canada Goose

Left: Giant Hogweed

Above: Himalayan Balsam

Below: Japanese Knotweed

Rhododendron

Montbretia

Ruddy duck

Grey squirrel

Wild fuschia

Variegated Yellow Archangel

Wild figs

# *Triumph over adversity: a new era for urban lichens*

## M. R. D. Seaward

Department of Geography & Environmental Sciences, University of Bradford,
Bradford BD7 1DP
m.r.d.seaward@bradford.ac.uk

Although adapted to survive in a very wide variety of environments, lichens are highly sensitive to human interference, especially air pollution. Although several early botanists noted that lichens did not thrive in polluted atmospheres, it was William Nylander who suggested that lichens might be used as indicators of air quality: his studies in the Jardin du Luxembourg, Paris showed the decline in and later disappearance of its epiphytic flora (Nylander, 1866, 1896). More than 90 years were to elapse before epiphytic lichens returned to the Jardin du Luxembourg (Seaward & Letrouit-Galinou, 1991).

Before the 1950s, air pollution studies involving lichens were mainly concerned with casual observations on impoverished urban floras of western Europe and Scandinavia. Only a few of these studies incorporated quantitative data. Sernander's (1926) pioneering work on the lichen flora of Stockholm recognized three zones, namely desert, struggle and normal, a system adopted by others, such as Haugsja (1930) whose work on Oslo provided the first maps of species showing their inner limits and struggle/normal boundaries. As well as their value in synoptic analyses, such studies enabled comparisons to be made of the severity, extent and pattern of air pollution between cities.

Several influential papers in the 1950s provided a springboard for the future development of lichen-air pollution studies. The first (Jones, 1952) contained details of lichen floras of tree boles at sites in England where air pollution levels were known, and numerous more sophisticated bioindicational scales based on his rudimentary scale were developed. Prior to this, an 'air pollution hypothesis' had been framed to explain the paucity of lichens in urban areas and the gradual decline in the European lichen flora in general since the Industrial Revolution which began at the end of the eighteenth century. However, the constituents of air pollution responsible for these phenomena had not yet been determined; indeed, Jones (1952) and others had indicated smoke and/or soot deposition to be the main cause. Although sulphur dioxide is now recognized as being the major culprit affecting lichens, other pollutants are implicated: the synergistic (exacerbatory/buffering) effects of the constituents of complex pollution burdens in stressed urban environments affects lichen performance in ways that are not easy to interpret.

The 'air pollution hypothesis' was challenged by Rydzak (1954, 1956–59, 1969) and Beschel (1958) on the basis of their work on the lichen floras of small towns. Rydzak proposed his 'ecological (or drought) hypothesis', the zonation patterns seen in urban lichen floras being due to adverse micro-climatic conditions, particularly lowered humidity, and increased temperature and frequency of drying/wetting regimes. Numerous researchers have commented on errors in Rydzak's interpretations of urban lichen floras (e.g. Coppins, 1973), concluding that the 'drought hypothesis' cannot be substantiated by field and laboratory investigations and that air pollution, particularly by sulphur dioxide, can better explain changes which have occurred in both urban and rural lichen floras. However, in naturally-occurring arid or sub-arid climates, both air pollution and drought can affect lichen distribution, as demonstrated by Nimis (1985) in the case of Trieste. Furthermore, the potential interaction of global warming on polluted environments cannot be discounted in future impact assessments of lichen performance.

However, it was the work of Barkman (1958) and Skye (1958, 1968) which most influenced future research on lichens and air pollution; Barkman's (1958) monumental treatise on their phytosociology and ecology proved of immense value in establishing protocols for the description and measurement of epiphytic lichens, and Skye recognized sulphur dioxide as the over-riding factor affecting an urban lichen flora. Skye's (1968) detailed investigation of the Stockholm lichen flora provided a model for numerous subsequent studies; as well as adopting Barkman's (1958) techniques, he examined past and present floras, including cartographical comparisons of surveys by Sernander (1926) and Hoeg (1934), and evaluated the importance of various environmental factors, particularly city climate, air chemistry and substrata, on their distribution patterns.

Field investigations were also in progress at this time to determine the relationship between lichen performance and air pollution level, the major detrimental component of which had by now been identified as sulphur dioxide. Since pollution level decreases with distance, early attempts to express this relationship in terms of the lichen flora took the form of graphical interpretations of species diversity counts and/or ecological behaviour along transects from areas judged to be subject to high pollution loads to sites in less polluted areas. Fenton (1960) pioneered the transect approach and demonstrated that measurements of percentage cover of particular groups of epiphytic species could be taken as a reliable index of air quality. However, all too often such transects followed the most easily accessible and convenient routes for the urban researcher which may not necessarily provide representative sampling sites for the pollutant(s) under investigation.

The key paper on zonal mapping by Hawksworth and Rose (1970) provided a qualitative scale for estimating sulphur dioxide air pollution using epiphytic lichens on non-eutrophicated or eutrophicated bark. This scale has stood the test of time, proving of immense value in air pollution monitoring in the British Isles and forming the basis for scales elsewhere. Although used to monitor the extent or spread of air pollution, particularly sulphur dioxide, it is less effective in some areas experiencing reduced levels of, or qualitative changes in, pollution as a result of clean air legislation, etc. (Seaward, 1987; van Haluwyn & Lerond, 1988; Richardson, 1988).

A clearly defined negative relationship exists between species diversity and sulphur dioxide, when the latter is more-or-less stable or increasing in concentration, with anthropogenic factors obviously over-riding such factors as climate and topography (Seaward, 1976b). However, due to recent dramatic reductions in air pollution levels in many cities, species diversity counts and distribution patterns no longer demonstrate a clear relationship with sulphur dioxide. Although multivariate analyses have shown that other factors are involved in such circumstances (Seaward, 1993, Table 1), the destructive role of other atmospheric pollutants, and indeed additional factors, has been insufficiently appreciated due to the inadequacy of existing monitoring equipment. The synergistic effects of the sum total of factors involved will in all probability never be fully appreciated by means of multivariate analyses; however, a representative spectrum of implicated pollutants has been subjected to this type of analysis (Ammann *et al.*, 1987; Herzig & Urech, 1991) and detailed studies in the Grenoble area of France have shown the complexity and interaction of different pollutants and other factors affecting the distribution of urban lichens (Gombert *et al.*, 2004, 2005).

Earlier studies revealed that a few lichen species obviously tolerated anthropogenically-modified atmospheres, and the terms 'toxiphilous' and 'poleophilous' were coined to describe lichens found in polluted and urban environments respectively; these were subsequently replaced by 'toxitolerant' and 'poleotolerant', since rather than liking such stressful environments, lichens tolerate them, often succeeding there due to a lack of competition, a substratum with a capacity to buffer harmful pollutants, or an input of nitrogenous compounds (see below). To succeed in urban environments, lichens must survive constantly changing factors; some species appear to be sensitive to short-term extremes and others to more long-term enhanced pollution levels. Strategies of toxitolerant species adapting to the multiplicity of stressful factors operating in polluted environments vary considerably; from an air spora rich in a variety of propagules, only a few species are capable of surviving germination and development. Once established, such species are often aggressive, and, under conditions of amelioration, have proved to be highly competitive, to the exclusion of those species which would normally grow under such regimes: substrata are dominated by one or a very few strongly competitive species with a high reproductive capacity and a tenacious hold on the substrata they colonize. An example of this is the rise to dominance of *Lecanora conizaeoides* throughout Britain (Seaward & Hitch, 1982, map 64), and indeed much of Europe, over the second half of the twentieth century: atmospheric amelioration in the 1970s and 1980s, resulting from the implementation of clean air policies, failed to break its monopoly of many substrata; only in the last 15 years or so has the monotonous verdure created by this and a few other species coating tree trunks in urban areas been superseded by more diverse epiphytic lichen assemblages, and indeed *L. conizaeoides* is now becoming rare or absent in many parts of Britain (Coppins *et al.*, 2001).

Several lichen species are highly successful in areas subject to relatively high levels of air pollution; their spread into the suburbs of British cities in recent years has been spectacular (Seaward & Henderson, 1984, 1991; Seaward *et al.*, 1994; Hawksworth &

McManus, 1989; Seaward, 1997). Although most of these are equally successful elsewhere, some which formerly had narrower ecological requirements and/or more restricted distributions appear to thrive on urban substrata subjected to new atmospheric regimes. From zonal studies in polluted areas it has been shown that the inner distributional limit of many lichens is quite clearly defined, as instanced by *Lecanora muralis* which has an ability to exploit man-made calcareous substrata (which counteract acid deposition), and thus extend its geographical and ecological ranges into urban areas (Seaward, 1976a, 1982). More recently, due to further reductions in air pollution levels, it has become highly successful in colonizing asphalt of urban streets and pathways. The ecological factors operating at distributional limits on or immediately preceding a particular date are critical for the lichen's performance or even for its existence. When investigating incipient changes in inner distributional limits, note should be made of which species survive in micro-climatic niches by chance establishment, which are relicts, and which appear to have responded to a change in atmospheric pollution.

Enlargement of the 'desert' for a particular species around a pollution source (or sources) clearly indicates rising levels of air pollution; conversely, decreasing pollution levels reduce the size of the 'desert'. However, interpretation of events through modelling derived from reasonably long-range field data of zonal changes has not necessarily been straightforward. Work on the lichen flora of the West Yorkshire conurbation in northern England over the past 38 years has shown that during the 1970s, following dramatic reductions in sulphur dioxide levels due to the implementation of clean air legislation, the 'desert' for several species actually enlarged: field data revealed that minor increases in sulphur dioxide at inner distributional limits exceeded the extinction threshold for some species (Seaward, 1993, Figure 2); by the early 1980s there was no change in the size of the 'desert', but from the mid-1980s onwards widespread reductions in pollution throughout both urban and suburban areas reduced the size of the 'desert'. This work clearly demonstrated that emission dilution, adopted by many industries and authorities as a solution to local pollution, can often have profound effects on the lichen flora beyond the local limits.

Biological scales for estimating sulphur dioxide air pollution (e.g. Hawksworth & Rose, 1970) are logarithmic in nature (Figure 1): major reductions in the level of the pollutant ($A^1$) at the polluted end of the scale may have little, or indeed no, effect on lichen species diversity ($B^1$); however, in a less polluted situation, the effect on lichen vegetation increases exponentially, such that only a minor increase in sulphur dioxide ($A^2$) would dramatically reduce species diversity ($B^2$). Furthermore, it should be noted that this model can be utilised to demonstrate the relationship between various pollutants and species diversity of different organisms not only spatially, as in the case of a transect to/from pollution source(s), but also temporally, as in the monitoring of a rise or fall in the pollution burden.

Although lichens respond relatively quickly to a rise in sulphur dioxide concentration and can therefore be satisfactorily employed to monitor ambient levels, in areas experiencing a fall in concentration their use as bioindicators is less straightforward and it is often difficult to demonstrate any relationship between species diversity and air pollution level (Seaward, 1976b).

The re-establishment of lichens in urban areas adopting clean air policies can be dramatic over a relatively short time-scale, but their recovery is not immediate, necessitating a time-lag, modelled in one instance to be about five years (Henderson-Sellers & Seaward, 1979); the rate of reinvasion varies considerably, the establishment of invading lichen propagules being influenced by a complexity of factors, particularly the nature of the substratum; older trees are more slowly recolonized than younger trees (Hawksworth & McManus, 1989) and certain common foliose species appear to be more successful in recolonization than crustose species. Nevertheless, a species diversity count, based on field records of frequently occurring lichens within a defined area, and not restricted to microhabitats, can be useful for evaluating the general level of air pollution, particularly in respect of transects radiating from urban/industrial complexes; furthermore, amelioration of the atmosphere of such areas can be effectively demonstrated by comparing transect data (Seaward, 1997, Figure 2).

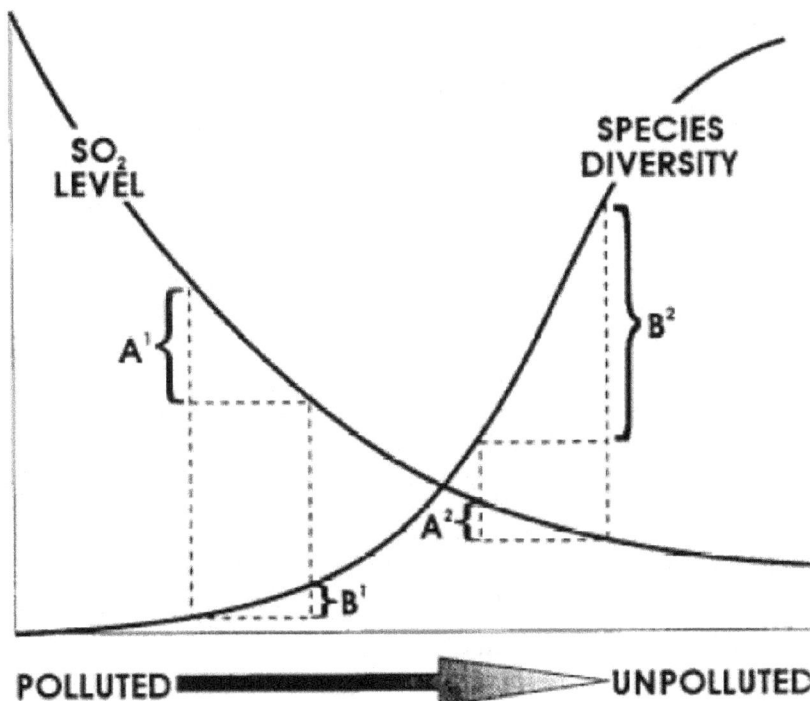

Figure 1. Model showing the relationship between species diversity and $SO_2$ level (Seaward, 1993). Biological scales are logarithmic in nature: major reductions in $SO_2$ ($A^1$) at the more polluted end of the scale will have little effect on biodiversity ($B^1$), but at the other end of the scale a minor increase in $SO_2$ ($A^2$) would dramatically reduce biodiversity ($B^2$). Note that the horizontal axis can be interpreted either spatially or temporally.

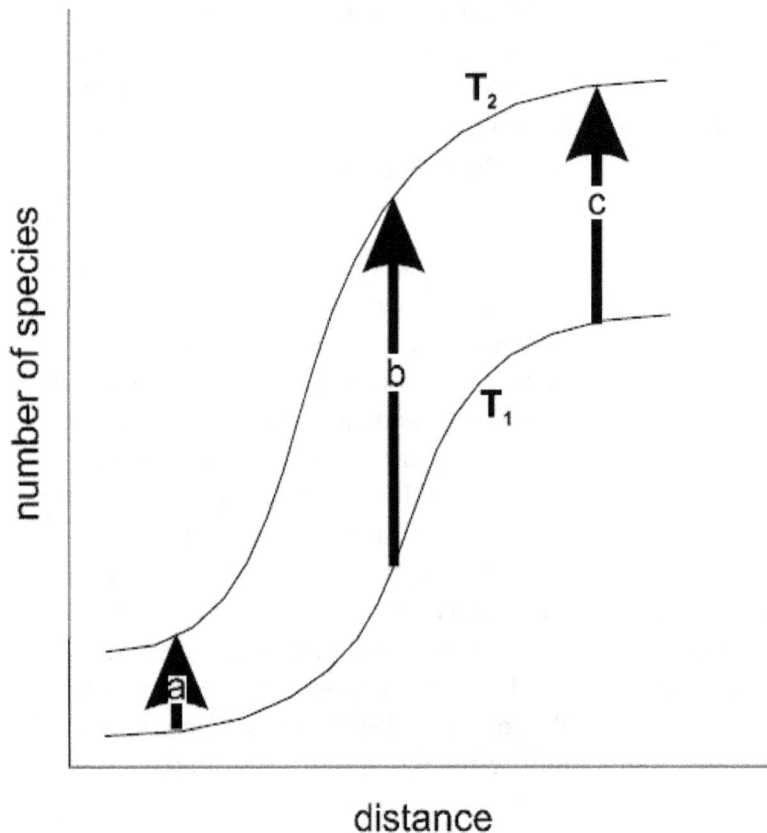

**Figure 2. Model derived from long-term field data (1972–1996) detailed in Seaward (1997, Figure 2) to show relative lichen improvement over time $T^1$ to $T^2$ in three urban areas in the following sequence: inner suburbia (b) > outer suburbia (c) > city centre (a).**

The West Yorkshire conurbation is one of the most intensively worked urban lichen floras, baseline data derived in the initial study from 1968 to 1975 (Seaward, 1975) being continually re-evaluated (Seaward, 1978, 1981; Seaward & Henderson, 1984, 1991; Seaward *et al.*, 1994) to determine in detail the distributional and ecological responses of lichens to environmental changes, more particularly quantitative and qualitative changes in air pollution. The improvement in the conurbation's flora since 1972 as a result of atmospheric amelioration, mainly brought about by clean air legislation, has been dramatic: 20 km transects from the centre of the conurbation (Seaward, 1997, Figure 2) show increases in species diversity for all zones, but inner suburbs are quite clearly experiencing greater changes. A model (Figure 2) based on these long-term observations demonstrates how inner suburbs now provide a milieu for a rich and biodiverse lichen flora.

Not surprisingly, poleotolerant lichens exploit the multiplicity of habitats afforded by urban environments (Seaward, 1979): the wide range of substrata, industrial wasteland, introduced shrubs and trees, gardens, sewage farms, etc. are subjected to varying degrees of nutrient-enriched dusts. Over 200 years, *c.*364 taxa have been recorded from the West Yorkshire conurbation; of these, at least 30 are extinct, but more than 232, including a significant number of taxa which have extended their ecological and geographical ranges, have been recorded during the past four decades. The graphical interpretation (Figure 2) is based only on those species which are found at a significant number of sites and are present in zones beyond the innermost zone in which they occur. Since species numbers more akin to rural/semi-natural sites are now commonly found in ameliorating urban environments, straightforward biodiversity counts are much less effective for evaluating urban environmental quality.

The improvement in lichen floras during recent years has been demonstrated in many European, Scandinavian and North American cities (Henderson-Sellers & Seaward, 1979; Seaward, 1980, 1990, 1997; Rose & Hawksworth, 1981; Showman, 1981, 1990; Kandler & Poelt, 1984; Hawksworth & McManus, 1989; Bates *et al.*, 1990; Seaward & Letrouit-Galinou, 1991); but it is apparent that re-establishing floras differ from those formerly wiped out by air pollution. From our knowledge of the behaviour of lichens subjected to sulphur dioxide pollution, it is possible to rank species according to their sensitivity to it, but variations in replacement assemblages necessitate modifications of bioindicational scales to account for these hierarchical changes.

The re-establishment of pendulous lichens such as *Usnea subfloridana*, *Evernia prunastri* and *Ramalina farinacea*, usually on *Fraxinus* and *Salix*, at numerous sites, including city suburbs, throughout Britain in the early 1980s reflected not only a decrease in air pollution level, but also a shift from dry to wet deposition of sulphur dioxide. It was observed that certain foliose lichens tolerant of a moderate level of sulphur dioxide had succumbed to a new atmospheric regime (i.e. 'acid rain'), the original assemblage being replaced by one which frequently included a few thalli of fruticose lichens known to be less tolerant of dry deposition sulphur dioxide. However, the stability of these taxa at many sites proved to be tenuous, since thalli only survived for one or two years. 'Acid rain' can affect lichens directly and/or indirectly by the acidification of substrata (Barkman, 1958; Skye, 1968; Grodzinska, 1971; Johnsen & Søchting, 1973). Other demonstrable effects, such as the development of epiphytic floras typical of more acid substrata, have been reported as a result of changes in the nature of pollutant emissions; since these floras include small species which are difficult to identify, particularly in the field, their distributions in urban areas have been difficult to determine, and they have only limited use as monitors of acid rain. Furthermore, deposition of nitrates via acid rain can produce a fertilization effect, as yet little studied, which encourages many lichen species to flourish, at least over the short term. Studies in Austria (Grill *et al.*, 1989) have attributed the decline of lichen floras in some city centres to increased motor vehicle traffic, the size of the lichen desert in Innsbruck, for example, increasing 5-fold between 1977 and 1987. However, field observations, supported by fumigation experiments, strongly suggest that there can

be a 'fertilization' effect on a lichen flora (see below), which is more pronounced when the sulphur dioxide burden is reduced.

Numerous species now spreading into suburbs are equally successful elsewhere but formerly had narrower ecological requirements and/or more restricted distributions. They are probably favoured by reduced competition, but urban and industrial dusts (mainly alkaline), soil fertilizers and other agrochemicals can promote unusual lichen floras. Thus, dust impregnated tree bark supports assemblages more commonly associated with saxicolous habitats (Gilbert, 1992); for example, the presence of *Candelariella vitellina*, *Lecanora dispersa*, *L. muralis*, *Phaeophyscia orbicularis*, *Physcia* spp., *Rinodina gennarii* and *Xanthoria* spp., particularly about the bases of trees in urban parklands, is in many cases a response to this phenomenon.

The recent spread of some species into city suburbs and elsewhere (e.g. *Ramalina farinacea*, *Xanthoria polycarpa*) may also be promoted by wind-borne nutrient-enriched dusts. In many areas of Europe, recent changes in lichen composition often reflect increased eutrophication as much as a reduction in sulphur dioxide levels. The spectacular decline in the lichen flora of The Netherlands between 1900 and 1974 (de Wit, 1976) was matched by changes in recent years, most notably an increase in nitrophytes resulting from increased ammonia emissions from intensive cattle husbandry (e.g. van Dobben & de Bakker, 1996; van Herk, 1999); the relationship between ammonia concentration, bark pH and composition of the lichen flora at sites subjected to such emissions was evident (de Bakker, 1989). Furthermore, Catteloin and Arvy (1989) in a bioindication study of Chinon, France, determined four pollution zones with characteristic lichen floras, but no pollution gradients, concluding that in some areas agricultural practices rather than industrial activities were more determinative of the lichen flora. According to Randlane and Trass (1991), nitrogen oxides and calcareous dust have lessened the lethal influence of sulphur dioxide on *Xanthoria parietina* in northeast Estonia. Similarly, a 'fertilization' effect generated from an agrochemical factory's emissions ameliorated a sulphur dioxide problem (Seaward, 1990), and its eventual closure adversely affected the local lichen flora. The term 'eutrophication' does not do justice to the excessive levels of nitrogen and other compounds being experienced by lichens; hence the term 'hypertrophication' has been applied to the interpretation of the resultant assemblages (Seaward & Coppins, 2004). The effects on lichens of an increase in nitrogen to the air pollution burden via vehicle emissions have yet to be fully investigated.

It is important to note that once sulphur dioxide air pollution, the governing factor for the reinvasion of lichens, has been suppressed/ eliminated, the ecological complexity of urban environments, in terms of the wide diversity of substrata, niches and habitats available, and the factors obtaining thereon, provides enormous possibilities. The wide range of available urban habitats now subjected to new atmospheric burdens, including hypertrophication, can be exploited by lichens to great advantage, as recently witnessed in many of our cities (Seaward, 1997; Seaward & Coppins, 2004).

The use of corroborative information obtained from bioindication surveys to support the limited data derived from physico-chemical gauges is strongly recommended. There is a continuing need for intensive lichen monitoring as a necessary component of any programme aimed at effective long-term observation of air pollution. However, any successful use of lichen monitors must employ strict methodological protocols, including a rigorous definition of the monitoring scales adopted. A complete analysis should be made of all the environmental parameters affecting the propagation, establishment and subsequent growth of the lichens under investigation. The use of bioindicators has the obvious advantage of permitting long-term monitoring without widespread establishment and maintenance of costly and sophisticated equipment. Unfortunately, most bioindication scales rely on species diversity counts, or at least on a fairly detailed understanding of lichen taxonomy. This makes the techniques as difficult to use as sophisticated physico-chemical measuring equipment, unless lichenologists train a new generation of young scientists to continue the solid groundwork that is now in place.

# References

Ammann, K., Herzig, G. R., Liebendorfer, L. & Urech, M. (1987) Multivariate correlation of deposition data of 8 different air pollutants to lichen data in a small town in Switzerland. In: *Advances in Aerobiology*: 401–406. Birkhauser Verlag, Basel.

Barkman, J.J. (1958) *Phytosociology and Ecology of Cryptogamic Epiphytes*. Van Gorcum, Assen.

Bates, J.W., Bell, J.N.B. & Farmer, M. (1990) Epiphyte recolonisation of oaks along a gradient of air pollution in south-east England, 1979–1990. *Environ.Pollut.* **68**: 81–99.

Beschel, R.E. (1958) Flechtenvereine der Stadte, Stadtflechten und ihr Wachstum. *Ber. Naturw.-med.Ver.Innsbruck* **52**: 1–158.

Catteloin, C. & Arvy, M.P. (1989) Étude cartographique de la flore lichénique du site de Chinon, Indre-et-Loire. *Acta Oecologia Applicata* **10**: 241–258.

Coppins, B.J. (1973) The 'drought hypothesis'. In: Ferry, B.W., Baddeley, M.S. & Hawksworth, D.L. (eds) *Air Pollution and Lichens*, 124–142. Athlone Press, London.

Coppins, B.J., Hawksworth, D.L. & Rose, F. (2001) Lichens. In: Hawksworth, D.L. (ed.) *The Changing Wildlife of Great Britain and Ireland*, 126–147. Taylor & Francis, London.

de Bakker, A.J. (1989) Effects of ammonia emission on epiphytic lichen vegetation. *Acta Bot.Neerl.* 38: 337–342.

de Wit, T. (1976) Epiphytic lichens and air

pollution in The Netherlands. *Bibl.Lichenol.* **5**: 1–115.

Fenton, A.F. (1960) Lichens as indicators of atmospheric pollution. *Ir.Nat.J.* **13**: 153–159.

Gilbert, O.L. (1992) Lichen reinvasion with declining air pollution. In: *Bryophytes and Lichens in a Changing Environment* (Bates, J.W. & Farmer, A.W. eds): 159–177. Clarendon Press, Oxford.

Gombert, S., Asta, J. & Seaward, M.R.D. (2004) Assessment of lichen diversity by index of atmospheric purity (IAP), index of human impact (IHI) and other environmental factors in an urban area (Grenoble, south-east France). *Sci.Total.Environ.* **324**: 183–199.

Gombert, S., Asta, J. & Seaward, M.R.D. (2005) The use of autecological and environmental parameters for establishing the status of lichen vegetation in a baseline study for a long-term monitoring survey. *Environ. Pollut.* **135**: 501–514.

Grill, D., Hafellner, J., Kaschnitz, A. & Pongratz, W. (1988) Neuer Erhebung der epiphytischen Flechtenvegetation in *Graz. Mitteil.Naturw.Ver.Steiermark* **118**: 145–155.

Grodzinska, K. (1971) Acidification of tree bark as a measure of air pollution in southern Poland. *Bull.Acad.Polon.Sci., ser.Sci.*, Biol.II **19**: 189–195.

Haugsja, P.K. (1930) Uber den Einfluss der Stadt Oslo auf die Flechtenvegetation der Baume. *Nyt Mag.Naturvid.* **68**: 1–116.

Hawksworth, D.L. & McManus, P.M. (1989) Lichen recolonization in London under conditions of rapidly falling sulphur dioxide levels, and the concept of zone skipping. *Bot.J.Linn.Soc.* **100**: 99–109.

Hawksworth, D.L. & Rose, F. (1970) Qualitative scale for estimating sulphur dioxide air pollution in England and Wales using epiphytic lichens. *Nature* **227**: 145–148.

Henderson-Sellers, A. & Seaward, M.R.D.

(1979) Monitoring lichen reinvasion of ameliorating environments. *Environ.Pollut.* **19**: 207–213.

Herzig, R. & Urech, M. (1991) Flechten als Bioindikatoren. *Biblioth.Lichenol.* **43**: 1–283.

Hoeg, O.A. (1934) Zur Flechtenflora von Stockholm. *Nyt Mag.Naturvid.* **75**: 129–136. Johnsen, I. & Søchting, U. (1973) Influence of air pollution on the epiphytic lichen vegetation and bark properties of deciduous trees in the Copenhagen area. *Oikos* **24**: 344–351.

Jones, E.W. (1952) Some observations on the lichen flora of tree boles, with special reference to the effect of smoke. *Revue bryol. lichénol.* **21**: 96–115.

Kandler, O. & Poelt, J. (1984) Wiederbesiedlung der Innenstadt von München durch Flechten. *Naturwiss.Rundsch.* **37**: 90–95.

Nimis, P.L. (1985) Urban lichen studies in Italy. I: The town of Trieste. *Studia Geobot.* **5**: 49–74.

Nylander, W. (1866) Les lichens du Jardin du Luxembourg. *Bull.Soc.bot.Fr.* **13**: 364–372.

Nylander, W. (1896) *Les Lichens des Environs de Paris*. Schmidt, Paris.

Randlane, T. & Trass, H. (1991) Kel silma samblike jaoks. 6. Lakmusliigid. *Eesti Loodus* **4**: 220–223.

Richardson, D.H.S. (1988) Understanding the pollution sensitivity of lichens. *Bot.J.Linn.Soc.* **96**: 31–43.

Richardson, D.H.S. (1992) *Pollution Monitoring with Lichens*. Richmond Publishing, Slough.

Rose, C.I. & Hawksworth, D.L. (1981) Lichen recolonization in London's cleaner air. *Nature* **289**: 289–292.

Rydzak, J. (1954) Rozmierszczenie i ekologia porostow miasta Lublina. *Annls Univ.Mariae Curie-Sklodowska*, C, **8**: 233–356.

Rydzak, J. (1956-59) Wplyw malych miast na flore porostow. *Annls Univ.Mariae Curie-Sklodowska,* C, **10**: 1–32, 33–66, 157–175, 321–398; **11**: 25–50, 51–72; **13**: 275–323.

Rydzak, J. (1969) Lichens as indicators of the ecological conditions of the habitat. *Annls Univ.Mariae Curie- Sklodowska*, C, **23**: 131–164.

Seaward, M.R.D. (1975) Lichen flora of the West Yorkshire conurbation. *Proc.Leeds Phil. Lit.Soc.,* Sci.Sect. **10**: 141–208.

Seaward, M.R.D. (1976a) Performance of *Lecanora muralis* in an urban environment. In: Brown, D.H., Hawksworth, D.L. & Bailey, R.D. (eds), *Lichenology: Progress and Problems,* 323–357. Academic Press, London.

Seaward, M.R.D. (1976b) Lichens in air polluted environments: multivariate analysis of the factors involved. In: Karenlampi, L. (ed.), *Proceedings of the Kuopio Meeting on Plant Damages Caused by Air Pollution,* 57–63. University of Kuopio, Kuopio.

Seaward, M.R.D. (1978) Lichen flora of the West Yorkshire conurbation - supplement I (1975–1977). *Naturalist* **103**: 69–76.

Seaward, M.R.D. (1979) Lower plants and the urban landscape. *Urban Ecology* **4**, 217–225.

Seaward, M.R.D. (1980) The use of lichens as bioindicators of ameliorating environments. In: Schubert, R. & Schuh, J. (eds), *Bioindikation auf der Ebene der Individuen,* 17–23. Martin-Luther-Universität, Halle-Wittenberg.

Seaward, M.R.D. (1981) Lichen flora of the West Yorkshire conurbation - supplement II (1978–1980). Naturalist 106: 89-95.

Seaward, M.R.D. (1982) Lichen ecology of changing urban environments. In: Bornkamm, R., Lee, J.A. & Seaward, M.R.D. (eds), *Urban Ecology,* 181–189. Blackwells, Oxford.

Seaward, M.R.D. (1987) Effects of quantitative and qualitative changes in air pollution on the ecological and geographical performance of lichens. In: Hutchinson, T. & Meema, K.M. (eds), *The Effects of Atmospheric Pollutants on Forests, Wetlands and Agricultural Ecosystems,* 439–450. Springer Verlag, Berlin.

Seaward, M.R.D. (1990) The lichen flora of Teesside. *Naturalist* **115**: 73–79.

Seaward, M.R.D. (1993) Lichens and sulphur dioxide pollution: field studies. *Environ.Rev.* **1**: 73–91.

Seaward, M.R.D. (1997) Urban deserts bloom: a lichen renaissance. *Bibl.Lichenol.* **67**: 297–309.

Seaward, M.R.D. & Coppins, B.J. (2004) Lichens and hypertrophication. *Bibl.Lichenol.* **88**: 561–572.

Seaward, M.R.D. & Henderson, A. (1984) Lichen flora of the West Yorkshire conurbation – supplement III (1981-83). *Naturalist* **109**: 61–65.

Seaward, M.R.D. & Henderson, A. (1991) Lichen flora of the West Yorkshire conurbation – supplement IV (1984-90). *Naturalist* **116**: 17–20.

Seaward, M.R.D., Henderson, A. & Earland-Bennett, P.M. (1994) Lichen flora of the West Yorkshire conurbation - supplement V (1991-93). *Naturalist* **119**: 57–60.

Seaward, M.R.D. & Hitch, C.J.B., eds (1982) *Atlas of the Lichens of the British Isles.* Volume 1. Institute of Terrestrial Ecology, Cambridge.

Seaward, M.R.D. & Letrouit-Galinou, M.-A. (1991) Lichen recolonization of trees in the Jardin du Luxembourg, Paris. *Lichenologist* **23**: 181–186.

Sernander, R. (1926) *Stockholms Natur.* Almquist & Wiksells, Uppsala.

Showman, R.E. (1981) Lichen recolonization following air quality improvement. *Bryologist*

**84**: 492–497.

Showman, R.E. (1990) Lichen recolonization in the upper Ohio River valley. *Bryologist* **93**: 427–428.

Skye, E. (1958) Luftfororeningars inverkan pa busk- och blad- lavfloran kring skifferoljeverket i Narkes Kvarntorp. *Svensk bot.Tidskr.* **52**: 133–190.

Skye, E. (1968) Lichens and air pollution. *Acta phytogeogr.suec.* **52**: 1–123.

van Dobben, H.F. & de Bakker, A.J. (1990) Re-mapping epiphytic lichen biodiversity in The Netherlands: effects of decreasing $SO_2$ and increasing $NH_3$. *Acta Bot.Neerl.* **45**: 55–71.

van Haluwyn, C. & Lerond, M. (1988) Lichénosociologie et qualité de l'air: protocole operatoire et limites. *Cryptog., Bryol.Lichénol.* **9**: 313–336.

van Herk, C.M. (1999) Mapping ammonia pollution with epiphytic lichens in The Netherlands. *Lichenologist* **31**: 9–20.

THIS PAPER WAS SUBMITTED IN 2005 AS PART OF AN URBAN ENVIRONMENTS CONFERENCE HELD IN SHEFFIELD.

# Relict communities and urban commons – urban distinctiveness, history and sustainable urban biodiversity

## Ian D Rotherham
Sheffield Hallam University

## Abstract

Urbanisation has increased and will continue to increase both in Europe and worldwide (The World Commission on Environment and Development, 1987). The importance of urban areas and the urban environmental footprint to nature conservation is therefore of great interest (Adams & Dove, 1989; Sukopp & Hejny, 1990; Sukopp, Numata & Huber, 1995).

Images of urban environments may reflect key periods of history such as Victorian England when industrial towns like Sheffield increased rapidly to become great cities.

**Sheffield's growing population: 1617: 2,207 people; 1700: 5–6,000 people; 1736: 9,696 people; 1801: 31,314 people; 1893: 333,922 people; in part through births, in part through migration, in part through urban spread**

Today the process continues in different ways and often in different places – but bigger and faster.

In understanding the environmental functioning and potential of this resource it is important to more fully appreciate the juxtaposition of people and nature through time. Unfortunately this is often lacking in site and strategic assessments. The need is for the often separate disciplines – ecology, earth sciences, archaeology, history, geography to be combined to gain a fuller vision of the resource, its past and its potential in the future. It seems to many that research funding initiatives such as URGENT, a major research council grant aid package in the early 2000s, fail to grasp this wider remit.

Central to understanding the ecology of urbanising areas is the interaction between old and new; between encapsulated remnants and establishing urban commons and ruderals. Ideas and concepts such as that of *Recombinant Ecology* (e.g. Barker (ed.), 2000) are developing in part to address the challenges of new communities and ecologies that emerge through the complex interactions of environmental change, history and culture.

In broader conservation terms too, urban areas and their impacts – environmental, social and economic – become increasingly important. It is the urban community and the urban politician that more and more determine the agendas and policies for the wider environment.

It education and awareness raising, urban areas are a show case not just for their own unique interests, but also for demonstrating to the town-dweller, the nature of the landscape beyond (Rotherham, 1999). These are cultural landscapes with an increasingly cultural ecology.

## Introduction

Urbanised landscapes present environments with patchworks of often fragmented, isolated habitat-types, but with a varied range of plant and animal species (Gilbert, 1989). Significant in the typical communities of urban commons are highly adaptive, native ruderal species of catholic habitat requirements and cosmopolitan occurrence (Hodgson, 1989). These occur with alien species; some broadly adapted to disturbed, productive sites, and others specialised to exploit more extreme environmental conditions of the urban environment (Gilbert, 1992a and b). The occurrence and potential of features such as relict heaths and marshes and the problems for their associated species were noted by Teagle in *The Endless Village* (1977), and the progressive degradation of encapsulated environments in the USA has been noted by Murphy (1988).

The diversity and interest of these urban commons communities may be ephemeral, but where relict landscapes are encapsulated and survive in urban zones, remnant communities of animal and plant species occur too. Ironically their survival here may be better than in the hostile zones of agro-industry beyond the city. However, the demise of relict or remnant species can also be linked to urban settlement and progressive urbanisation. This may be through direct impact and loss or through processes such as pollution-associated eutrophication, and what I have described as 'cultural severance'. Where pockets of native species-rich vegetation and fauna survive it is often along steep stream valleys such as Rock Creek Park in Washington D.C., in the USA (Murphy, 1988), or the major rivers and valleys in cities like Sheffield, UK (Bownes *et al.*, 1991; Rotherham, 1999). Such relicts are generally typical constituents of communities from semi-natural, unimproved environments, and may occur as '*habitat patches*' of woods, heaths, wetlands and grasslands.

Such relict communities or community fragments in urban landscapes may offer long-term, sustainable environments of regional significance for nature conservation. They are relatively stable and robust; more so if appropriate site management can be achieved. With on-going debates on environmental sustainability, *biodiversity* and the potential value of urban areas for biodiversity (nature??) sustainability (conservation??), and the environmental impacts of urbanisation, are key issues.

The influence of rural land-use pattern and ownerships on subsequent urban patterns are well-known to urban geographers such as Ward (1970) writing about Leeds. These

patterns and processes are not always fully appreciated by ecologists and others today. Yet the evidence is there if we look:

- Landscape
- Archaeology
- Ecology
- Soils and Sediments
- Land-use
- Place names
- Archives
- Memories

These relict communities in urban areas have conservation potential as pools of high and sustainable biodiversity; with enhanced value through accessibility to local people (Jones & Rotherham, 1998; Rotherham, 1994a). To understand them however, requires a multi-disciplinary approach to address the complex facets and issues.

Tracking the changes in landscape or the spread of a relict or alien plant for example requires an awareness of many things other than pure ecology. Furthermore, the understanding gained can significantly change our perception of an area or of an environmental trend.

The urban footprint extends far beyond the administrative boundaries of any individual town or city – through social and economic processes as well as ecological. It occurs increasingly in and around settlements from the smallest to the largest but in differing degrees. The processes are complex and may extend across and through the landscape along routeways – roads, rail, canals, etc. In use or abandoned these may form linear wildlife habitats and green corridors. Written across the CANVAS of the landscape they may also present as archaeology and history.

We can track the spread of alien species out into the hinterland, and now the return movement of animals such as badger and deer back into the urban zone. Where these linear features have crossed relict areas in the landscape – woodland fragments, ancient hedgerows – species such as bluebell and dog's mercury may hitch a ride along the conduit to new opportunities.

**Urban environments influence biodiversity conservation through:**

- **The impacts of urban centres on regional environments and on the biodiversity and sustainability of surrounding areas**
- **Their contributions to the maintenance of biodiversity at local and regional levels**
- **The accessibility of urban areas to an increasingly large proportion of the population – including politicians and other decision-makers who influence economics and politics at regional and national levels**

These issues are discussed by Rotherham (1996, 1999) and Jones & Rotherham (1998). Tylka *et al.* (1987), noted twin goals for wildlife management and nature conservation in urban areas:

1. **The maintenance of regional species composition and abundance (regional biodiversity)**

2. **The provision of opportunities for human contact with nature**

## Travels through time: relict landscapes in the city

Abandoned countryside locked in the urban zone offers unique wildlife and social interest; *'time-capsules'* of former landscapes and land-use. These historic landscapes present a resource of relict communities (Rotherham, 1994b and 1996) with plants and animals as *'indicators'* of antiquity or continuity through time. Presence in an urban ecosystem may be clear evidence of connectivity with the former rural environment; perhaps removed in time by decades or centuries.

Urban landscapes offer serious challenges to many species. Isolation and limited mobility can be an issue. Over time however this may be more cosmetic than real for some wildlife species, able to '*hop*' or '*hitch*' from one '*habitat-node*' to another.

These encapsulated landscapes in urban and rural-fringe areas are not untouched, pristine environments; far from it. Location in heavily populated areas causes problems and generates distinctiveness. Importantly for conservation these areas give the urban community access to, and an insight into, the unique conservation value of their own historic landscapes. This is a far more meaningful experience than that provided by nature conservation sites in the remote countryside. These urban places are human-influenced with distinctive local character and relevance (Rotherham, 1991).

They are often intimately mixed with post-industrial and urban commons environments, as part of the '*hide-and-seek landscapes*' of John Box (1995).

# Discussion

## Urban Commons and Relict Environments

These historic environments may also have a unique and irreplaceable contribution to the sustainable conservation of local and regional biodiversity. They have a strong degree of local character and distinctiveness, reflecting both the environmental conditions of an area, and human exploitation. Once lost, their very individual lineage cannot be re-created.

Many of these areas are under serious threat of loss or degradation; often due to neglect, inappropriate management, theft of water catchment (leading to long-term, ecological drought), atmospheric pollution fall-out (nitrates, etc.) and associated eutrophication. The functioning of these integrated processes in the full context of their historic identity requires urgent investigation, but this is a neglected field.

Importantly, some progress has been made in terms of large-scale experiments in reintroducing sympathetic management to the relict areas (utilising grant-aid schemes such as Countryside Stewardship) with remarkable results in some cases. The substantial degree of site recovery, and the re-appearance of key fauna and flora on sites, especially species

believed to be locally extinct (such as in the Gleadless Valley and at Woodhouse Washlands) is very encouraging.

## Urban commons and encapsulated relicts

### Non-industrial commons

Gilbert (1989, 1992) focused attention on the special local character and relevance of the 'urban commons'. But these early successional communities especially if they follow non-industrial uses are often

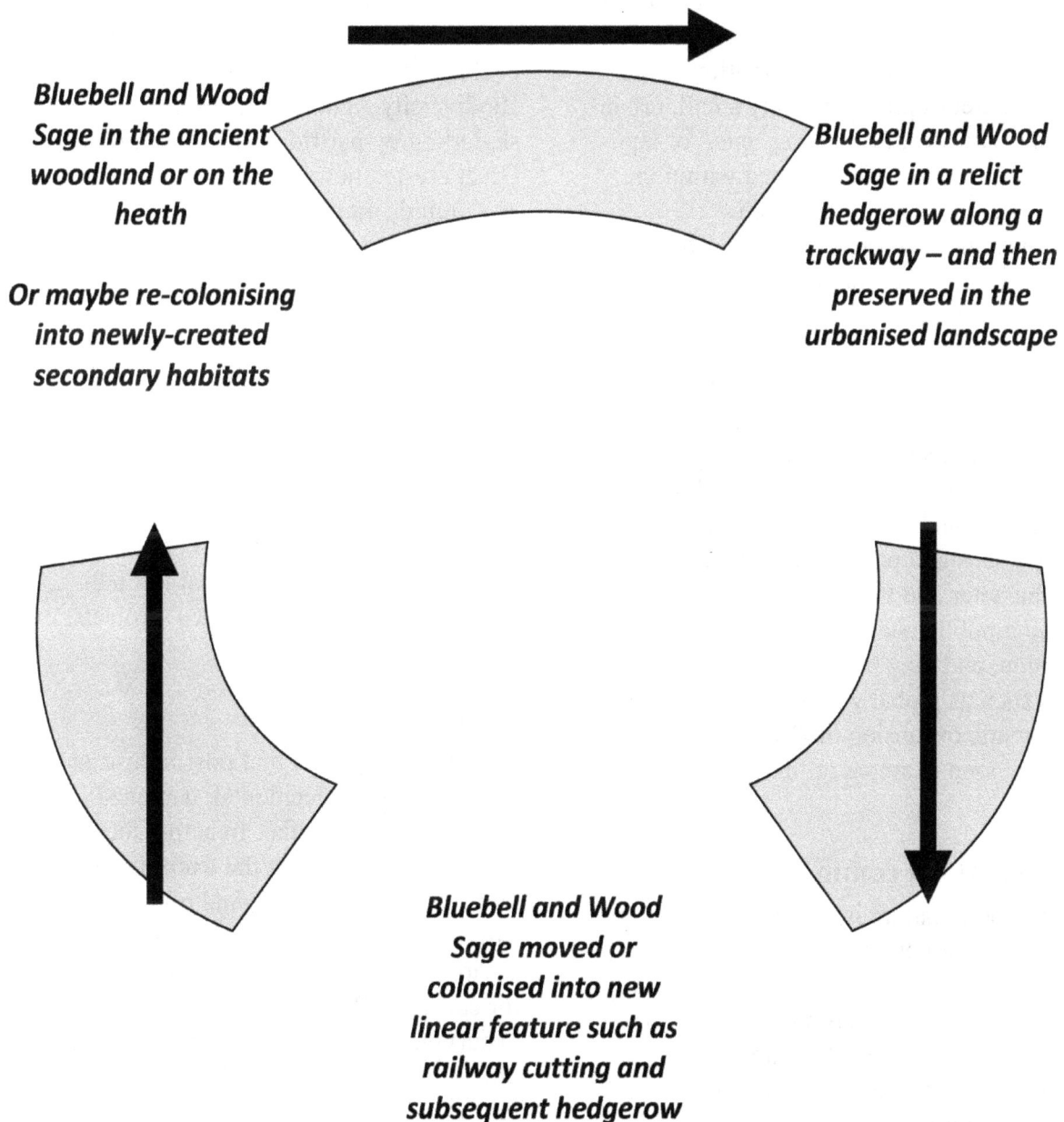

*Bluebell and Wood Sage in the ancient woodland or on the heath*

*Or maybe re-colonising into newly-created secondary habitats*

*Bluebell and Wood Sage in a relict hedgerow along a trackway – and then preserved in the urbanised landscape*

*Bluebell and Wood Sage moved or colonised into new linear feature such as railway cutting and subsequent hedgerow*

**Figure 1: a simplified schematic suggestion of relict habitat plants moving within the urbanising landscape.**

ephemeral; powered by disturbance and high levels of nutrients. They may be of strongly local character in their early years, but declining in interest soon after. True, the longer-term, interest of later successional stages, perhaps distinctive and spontaneous urban woodlands with intimate admixtures of native and exotic species, may be of great interest. Unfortunately, many urban commons sites are '*recycled*' into development, long before this can happen; the self-sown urban commons woods are often less valued than the planted impositions of planned restoration.

The smaller urban commons are inherently ephemeral in time and space, in ecological character and associated conservation value. The typical, eutrophic, relatively small, urban commons sites form a floating '*pool*' of land being recycled and re-distributed within the urban zone. This can be managed sympathetically and proactively to provide a sustainable but transient environment available to more mobile, ruderal species; colonisers hopping and hitching from one '*habitat node*' to the next, as sites are developed or succession moves on. Whilst new sites are created through development, these species will persist to be joined by ever more species and varieties that escape from urban gardens and parks. The latter reflect local character and the fluid tastes of the gardening public and the landscaping profession, and may be aided and abetted by factors such as global warming. The recent and dramatic expansion of *Buddleia* in urban areas is a good example of this process.

## Post-industrial commons

The ephemeral nature is not the case, and in particular some post-industrial sites, with extreme environmental conditions are both diverse and long-term sustainable (Shaw (1998) for example). Rotherham (1999) described Holbrook Heath, a site that demonstrates the potential interest of such areas. This same work also drew attention to the need for management if the special nature conservation interest is to be maintained beyond perhaps 60–70 years. At around 30–40 years after abandonment the rapid succession

of birch woodland is significantly affecting much of the site, and there total inertia in terms of any effective management action. Subjective perhaps, but it is tempting to imagine a prettier, more pristine site of lesser conservation significance would not be ignored in this way.

Work by Middleton (2000), Lunn (2001) and Rotherham (1999, 2003), suggests a further twist to the story. Many of these post-industrial sites are already of immense nature conservation value and it can also be argued that they possess huge cultural resonance and potential, along with in some cases, important and early industrial archaeology. Many post-coaling sites are often havens for National Biodiversity Action Plan species such as skylark, grey partridge and corn bunting. Great crested newts, dragonflies, water voles and orchids may be abundant in and around ponds and marshes accidentally left on abandonment of the industry.

Furthermore, important plant communities – heaths and grasslands in drier areas, and fen, marsh, and even bog in wet zones are also found and sometimes in abundance. These communities and species already contribute to the targets for local and national BAPs and yet are being accidentally lost through '*restoration*' and '*reclamation*' projects.

## Why is this?

It is often suggested that post-industrial or derelict sites are accidental, damaged and of little value or potential. In actual fact this assessment is far from the truth. Limited in terms of nutrient levels, and often with extreme physical and sometimes chemical environments, these areas offer opportunities for species of both animals and plants adapted to survive such difficult conditions. These are often rare species. They also provide habitats for the many species of conservation significance that have been squeezed out of the wider landscape by disturbance and by nutrient fallout and other contamination or eutrophication. A former colliery site at Holbrook near Killamarsh (Rotherham, 1999) is essentially abandoned and unmanaged. Yet it has over twenty species of *Cladonia* lichen, and over ten species of *Sphagnum* moss. For these alone it is the most significant site in the region. It is home to over twenty species of butterfly, and numerous other Red Data and BAP animals and plants – several of which are especially protected. (These include Badger (*Meles meles*), Harvest Mouse (*Micromys minutes*), Water Vole (*Arvicola terrestris*) and Great Crested Newt (*Triturus cristatus*)). Interestingly too, the site includes a coal spoil heap naturally regenerating to birch and oak woodland. The conditions here are extreme desiccation with extensive bare ground. This is now the only remaining such area in the district and is not only an interesting wildlife habitat, but also a unique piece of cultural and landscape archaeology too.

And yet despite this conservation groups such as the Sheffield Wildlife Trust in the 1990s were unable to support lobbies for the site's conservation, and it still remains under threat. It was not a priority and had no champion local community, and no '*soft funding*' attached. The response driven by economic necessity not nature conservation is understandable but the problem remains.

A consequence is that in the period of around fifteen years since the original research, much of the distinctiveness and importance of the site has gone as a consequence of rapid ecological, successional change. The trend has been towards species-poor birch wood, to red fen (itself of local interest) and to tall herb and scrub. A new, major routeway was cut through one of the main areas of species-rich, relict grassland.

## Replaceable?

It is assumed that these sites are inherently replaceable. They are not. Unlike many urban commons sites that have transient communities with species able to move between '*habitat islands*' as they become available, these post-industrial sites represent unique time-lines of accidental and

unrepeatable history. Many acquired species from the wider landscape at the time of their creation and development (see Rotherham, 1999). These are the same plants and animals now lost from the wider environment. They could not re-colonise a new derelict site even if it were created and the conditions suitable. These sites may hold examples, albeit impoverished, of the landscape from the early Industrial Revolution now largely lost or degraded in the wider countryside; an exciting discovery.

There is a further issue with the present climate of intervention conservation and suggestions to reintroduce species formerly lost (e.g. *Sheffield Biodiversity Action Plan*, 2002). For these conservation programmes to be successful it is necessary to have extensive and sustainable sites with the required environmental or management conditions in place. Without this they will fail, or will need conservation management for which there is no sustainable funding identified in the long-term. However, the unique histories and extreme conditions of many post-industrial sites, as well as their extensive nature make them eminently suited to the long-term survival of these animals and plants – and often without the need for reintroduction.

The recycling of land may be strongly episodic with cities like Sheffield subject major disruption and the creation of widespread urban commons twice during the twentieth century. First was during the Blitz of the early 1940s, and second was during the now on-going transformation from an industrial to a post-industrial city begun during the 1970s and 1980s. These influence both post-industrial and post-housing urban commons. There has been a third influence on the urban commons resource through the impacts of transport infrastructure, very neatly exemplified by the 1960s decline of the railways, and the creation of the so-called '*Beeching forest*' (Chris Baines, pers. comm.). These have their particular character as linear features and a value as '*green corridors*', although the best and most substantial examples in cities such as Sheffield were not valued sufficiently to save them from reintegration into the economic machinery of the region.

## Conclusions

With the exception of large-scale, post-industrial environments, the contribution of these urban commons sites to biodiversity conservation at the regional level will be limited. The species that tend to colonise are generally either catholic species with very cosmopolitan requirements, casual species of high mobility, or chance species exploiting ephemeral and bizarre environments. They are not usually the uncommon and threatened species genuinely native to the region and typical of the biodiversity of the '*natural area*' or of the local character of the area. They may be of great natural history interest, and of local, cultural value, but are only rarely of great conservation significance. Indeed the difficulties in raising awareness and generating any sort of substantial conservation lobby or effective site management, even for the most exciting of these sites, from local authorities or wildlife trusts and other organisations suggests that these limits will remain. The example of Petre Street in Sheffield demonstrates this very effectively. However, as long as the development process generates more sites then the species will move on. I suggest that the communities and the species mixes will continue to evolve and change.

Alongside encapsulated relicts the conservation potential of these secondary sites (accidental or deliberately created) should not be dismissed. A key factor in the potential importance of large-scale, post-industrial urban commons to local and regional biodiversity is the availability of relict sites from which re-colonisation by locally-native, rare species can occur. These sites are influenced by and reflect the character of their location, themselves becoming part of the urban historic landscape. At Holbrook Meadows in the Rother Valley, the juxtaposition of relict and secondary environments was the reason for the main ecological and conservation interest. This

potential was affected by the degree to which the site either included or was close by relict areas; or had generated through its industrial processes environmental conditions similar those of the '*ancient*' relicts.

Ancient, relict landscapes have historic and ecological links with a unique contribution to local and regional biodiversity inextricably associated with this lineage. Those remaining today are tiny vestiges of once widespread environments and originated through interactions of physical environment and social and economic function over time. Isolated and removed from these functions many sites decline through abandonment, inappropriate management, or other urban impacts. They may even be considered by some to be out of place in a modern urban context. However, when managed sympathetically they present relatively stable and sustainable urban ecosystems and are vital conservation resource.

Relict communities of wetlands, woodlands, grasslands and heaths have been shown to respond well to conservation management that mimics traditional, economically-functional land-use. Experiments conducted over a ten-year period, suggest that substantial recovery is possible.

Urban commons environments have a unique local history and relevance to local people, especially in their early successional stages and where extreme or bizarre conditions occur. They may be ecologically diverse and of great interest to local naturalists and often to local residents. Some sites may be particularly rich in relatively commonplace but attractive wildlife species - wild flowers, butterflies etc., and even hold local, regional and national rarities.

Some of the most exciting urban commons sites are those combining encapsulated relicts, acquired species from historic landscapes and the new cultural ecologies; but these are hard to absorb into the current planning system. As a culture we have a tendency towards the twin desires for control and for tidiness. Attempts to genuinely enthuse local people and planners to recognise and care for these areas are fraught with difficulties, and education at all levels is needed to change mind-sets that block effective action. These issues were eloquently expressed by Middleton (2000), with reference to the reclamation of former colliery sites in South Yorkshire, and the often total neglect (and hence destruction) of their nature conservation and cultural resource.

Most urban commons sites are transient and through process of ecological succession and economic re-development. They may converge rapidly to a bland uniformity, and are frequently either landscaped or developed before late successional species can take hold. This is not always the case, and large-scale, post-industrial sites may generate long-term sustainable communities of high conservation value and local cultural resonance. Long-term management of these larger post-industrial sites and short-term strategies for the floating pool of ephemeral sites may combine with positive conservation of relicts as a key to urban ecological sustainability.

To maintain and enhance interest in the longer-term management to perhaps mimic some of the traditional land management of the wider countryside may also be necessary (Jones & Rotherham, 1998). However, even before this point is reached there needs to be a wider recognition of the nature of the resource and its time context – both backwards to history to help inform our understanding, and forwards from the present to anticipate future changes. This poses a serious challenge to urban conservationists, to many planners, ecologists and archaeologists. The need is to understand the past, to inform the present and to influence the future. There is still a long way to go and many opportunities are being lost.

# References

Adams, L.W. & Dove, L.E. (1989) *Wildlife Reserves and Corridors in the Urban Environment.* National Institute for Urban Wildlife, Maryland, USA.

Anon. (1987) *Our Common Future.* The World Commission on Environment and Development. Oxford University Press, Oxford, UK.

Anon. (1994) *Biodiversity. The UK Action Plan.* HMSO, London.

Barker, G. (ed.) (2000) *Ecological recombination in urban areas: implications for nature conservation.* English Nature, Peterborough, 21–24.

Barker, G., Luniak, M., Trojan, P. &Zimny, H. (eds) (1994) Proceedings of the Second European Meeting of the International Network for Urban Ecology. *Memorabilia Zoologica,* **49**, Warsaw.

Bornkamm, R., Lee, J.A. and Seaward, M.R.D. (eds) (1982) *Urban Ecology.* Blackwell Scientific Publications, Oxford. Box, J. (1998) Post-industrial landscapes: challenge and opportunity. *Landscape Archaeology and Ecology,* **3**, 117–124.

Bownes, J. S., Riley, T., Rotherham, I. D. & Vincent, S.M. (1991) *Sheffield Nature Conservation Strategy,* Sheffield City Council, Sheffield.

Clarkson, K. & Garland, S.P. (1988) Colonisation of Sheffield's Urban Wasteland. Vascular Plants. *Sorby Record,* **25**, 5–21.

Fairhead, W., Baker, A., Rotherham, I.D. & Nowell, R. (1996) *An Ecological Assessment of Gleadless Valley.* Sheffield Centre for Ecology and Environmental Management, Sheffield.

Gilbert, O.L. (1989) *The Ecology of Urban Habitats.* Chapman and Hall. London.

Gilbert, O.L. (1992) *The flowering of the cities...The natural flora of 'urban commons'.*

English Nature, Peterborough, UK.

Hodgson, J.G. (1986) Commonness and Rarity in Plants with Special Reference to the Sheffield Flora. *Biological Conservation*, **36**(3), 199–252.

Jones, M. & Rotherham, I.D. (1998) 'Eyes have they but they see not': changing priorities and perceptions of countryside in urban areas. *Landscape Archaeology and Ecology*, **3**, 19–24.

Lazenby, A.S. (1983) Ground Beetles (Carabidae) and other Coleoptera on Demolition sites in Sheffield. *Sorby Record*, **21**, 39–51.

Middleton, P. (2000) The wildlife significance of a former colliery site in Yorkshire. *British Wildlife*, **11**(5), 333–339.

Murphy, D.D. (1988) *Challenges to biological diversity in urban areas.* In: *Biodiversity.* Wilson, E.O. (ed.) (1988). National Academic Press, Washington D.C., USA, 71–76.

Nilon, C.H. & Pais, R.C. (1997) Terrestrial vertebrates in urban ecosystems: developing hypotheses for the Gwynns Falls Watershed in Baltimore, Maryland. *Urban Ecosystems*, **1**(4), 247–257.

Rotherham, I. D. (1991) *Our neglected urban woodlands – a local authority perspective.* In '*A Seed in Time*', Proceedings of the Second UK Conference on Urban Forestry, 107–211.

Rotherham, I. D. (1994a) *The role of local authorities in conserving biodiversity through environmental management and land-use planning.* Proceedings of the VIth International Congress of Ecology, Manchester 1994, 77.

Rotherham, I. D. (1994b) Conserving wildlife in urban relict countryside, *Bulletin*, Yorkshire Naturalist Union, **22**, 2–4.

Rotherham, I.D. (1996) *Biodiversity in urban*

*areas, the contrasting values of relict and recently-created landscapes*, Proceedings of the Royal Geographical Society and The Institute of British Geographers Annual Conference, Strathclyde, January 1996.

Rotherham, I.D. (1998) If you go down to the woods today you're sure of a big surprise – they've gone! Comments on the progressive destruction of historic landscapes. *Landscape Archaeology and Ecology*, **3**, 145–151.

Rotherham, I.D. (1999) Urban Environmental History: the importance of relict communities in urban biodiversity conservation. *Journal of Practical Ecology and Conservation*, **3**(1), 3–22.

Rotherham, I.D. (2000) *People, perception and fashion.* In: Barker, G. (ed.): *Ecological recombination in urban areas: implications for nature conservation.* English Nature, Peterborough, 21–24.

Rotherham, I.D., Spode, F., and Fraser, D. (2003) Post–coalmining landscapes: an under-appreciated resource for wildlife, people and heritage. In Moore, H.M., Fox, H.R. and Elliot, S. (eds) *Land Reclamation: extending the boundaries.* Published: A.A. Balkema Publishers, Lisse. 93–99.

Rotherham, I.D. & Whiteley, D.W. (1995) *The Importance of Relict and Created Wetlands with reference to Invertebrates and Vegetation in Urban and Post-Industrial Sites. Preliminary Findings From The South Yorkshire Biodiversity Research Programme. British Ecological Society Conference: Recent Advances in Urban and Post-industrial Wildlife and Habitat Creation.* Leicester, 20–22, March, 1995. Poster Presentation.

Shaw, P.J.A. (1998) Conservation management of industrial wastes. *Practical Ecology and Conservation*, **2**(1), 13–18.

Sukopp, H. & Hejny, S. (eds) (1990) *Urban Ecology. Plants and Plant communities in urban environments.* SPB Academic Publishing. bv, The Hague, The Netherlands.

Sukopp, H., Numata, M. & Huber, A. (eds.)

(1995) *Urban Ecology as the basis of urban planning*. SPB Academic Publishing, The Hague, The Netherlands.

Teagle, W.G. (1977) *The Endless Village*. Nature Conservancy Council, Shrewsbury, UK.

Tylka, D.L., Schaefer, J.M. & Adams, L.W. (1987) *Guidelines for implementing urban wildlife programs under state conservation agency administration. In: Integrating man and nature in the metropolitan environment*. (Adams, L.W. & Leedy, D.L. (eds), National Institute for Urban Wildlife, Columbia, M.D., USA. 199–205.

Whiteley, D.W. (1988) Petre Street Urban Nature Zone. *Sorby Record*, **25**, 50–55.

Wild, M. & Gilbert, O.L. (1988) *Sheffield Inner City Habitat Survey*. Sheffield City Wildlife Group, Sheffield.

# Urban ecology and conservation: neglected and undervalued subjects?

## Annie Chipchase[1] and Peter Glaves[2]

Formerly: [1]London Wildlife Trust, Harling House, 47–51 Great Suffolk Street, London, SE1 0BS.
Formerly: [2]Department of Earth and Environmental Science, University of Greenwich, Chatham Maritime, Kent, ME4 4TB; now University of Northumbria.

## Abstract

Urban ecology and urban conservation have been around for as long as there have been ecologists and conservationists, although they remain the Cinderella's of their rural counterparts. This paper reviews the attitudes of ecologists and conservationists towards urban ecology and conservation. Attitudes which, with a few notable exceptions, are generally quite negative about urban areas and, indeed, may reflect a broader negative attitude towards the urban environment. Yet cities clearly contain areas of high conservation value, as demonstrated by the number of sites with nature conservation designations in cities such as London. Urban areas are important ecologically; they are also accessible to scientists and the general public. With ever increasing growth in urban populations and urban land cover, there is greater need to understand and, where appropriate, protect urban habitats. The paper argues that it is time we re-evaluated our urban environment. To facilitate this we need more ecologists and conservationists committed to the urban environment and dedicating more research to these areas.

## Introduction

The attitudes of many ecologists towards the environment in our cities are described by Sutherland (1999), as follows: 'urban landscapes [as] concrete jungles, paved wall-to-wall with tarmac – a hostile environment where nature struggles to survive'. Urbanisation has certainly created an environment epitomised by its man-made structures, its 'artificiality' standing in direct contrast to the 'natural world'. Yet the natural world is not absent from cities. This paper reviews the history of urban ecology and conservation, specifically looking at the attitudes that underlie the current and historical neglect of these subjects.

## General attitudes towards cities

Cities have always had a mixed press. Babel was associated with pride, Babylon with corruption, Sodom and Gomorrah with perversion, for Christians, the city represented both heaven and hell, *civitas Dei* and *civitas Diaboli* (Porter, 1996). Western attitudes over the last two hundred years have continued to demonstrate this ambiguity, e.g.

'Cities are the final pit of the human spirit' (Rousseau, n.d.).

'The town is the symbol of society ... of science, art, culture and religion. The country is the symbol of God's love and care for man' (Howard, 1898).

'When a man is tired of London he is tired of life' (Johnson, n.d.).

Many opinions about cities appear to derive from a perception that the city is the antithesis of the countryside (Adams, 1996). This duality sometime referred to as the 'second world', sees nature, and therefore ecology and conservation, as being apart from the city. The environment is something we visit at the weekend when we leave the city; it is somehow separate from our everyday lives. Today 89% of the population of England and Wales, including most conservationists and ecologists, live in urban areas covering just 7.7% of the land (Comedia Demos, 1995). By the year 2020 this will 'be true for most of the world' (Baines, 1999). Urban ecology is the ecology of the environment in which most people live, yet there has been a tendency among ecologists and naturalists to ignore urban habitats, considering them as being too far removed from nature (Freethy, 1986; Cairns, 1987; Botkin, 1990; Rees, 1997; Adams, 2003).

The main emphasis of the nature conservation movement since the late 1800s has, however, been concerned with the protection of semi-natural rural environments (Evans, 2003). Indeed, this is reflected in a number of histories and reviews of nature conservation, e.g. Stamp (1969) and Marren (2002) who largely ignore urban conservation. For example Sheail (1976) makes no reference to urban environments and Evans (2003) includes only three references to urban conservation. Unlike some of its European counterparts, the British National Vegetation Classification (Rodwell, 1991 to 2000 *et seq*) refers only indirectly to urban habits under its open vegetation communities. Some publications with a specific urban focus, in particular Nicolson-Lord (1987), do however, evaluate the roots and achievements of the urban conservation movement, and thus, acknowledge its existence. That aside, the cultural resonance of the rural environment, particularly for the English middle classes (Adams, 2003), and perhaps also for conservationists and ecologists, dates back even further and in many cases is much stronger. Thus, the study of urban conservation has remained rather neglected in favour of its rural counterpart (Cole, 1983).

There was renewed interest in urban ecology in the 1960s linked to the 'environmental movement' which, in the 1970s developed into a recognisably separate discipline (Nicholson-Lord, 1987). The coining of the term 'unofficial countryside' by Richard Mabey (1973) to describe some urban habitats afforded them their own distinct identity. The 1970s and 1980s saw the rise of the urban wildlife groups (Smyth, 1987) and of practical conservation guides (e.g. Flint 1985; Emery, 1986). Cole (1983) illustrated ecologists 'increasing awareness of nature in urban locations' when reviewing, on behalf of

the NCC, a number of inner city conservation projects from the early 1980's. Jenny Owen's garden survey between 1972 and 1986, just 3.8 km from Leicester's city centre also demonstrates how attitudes to ecology in the urban context have broadened (Good, 2000). This is reflected in the publication of general urban nature books such as the *Wild Side of Town* (Baines, 1986). In the early 1980s Operation Groundwork was set up to develop programmes which integrated socio-economic and ecological regeneration in inner city areas. Other urban initiatives include inner city farms, nature conservation areas in urban parks and schools and, more recently, the development of urban nature conservation strategies and biodiversity action plans.

Official recognition of the importance of nature conservation in urban areas followed the restructuring of the planning system as a result of the 1974 Local Government Act, which created the seven Metropolitan County Councils. The inability of the Nature Conservancy Council to provide information on the natural resources within the Metropolitan areas (Barker, 1996) led to 'the first comprehensive survey of the wildlife resource of an entire Metropolitan region' (Hornsby, 1990). The survey produced the first nature conservation strategy (West Midlands County Council, 1984), and the publication of *The Endless Village* (Teagle, 1978). Recognition of urban nature conservation resources and their relevance to the planning system was reflected in government circulars, e.g. *Nature Conservation and Planning* (DoE, 1977), and in the urban ecological planning guides produced by English Nature in the 1980s (e.g. Welsh, 1989). The legal requirements to consider nature in planning as part of environmental impact assessments may have stimulated this interest.

Over the last 20 years, there has been a development in urban-based research. This is probably due, in part, to issues relating to accelerating human population growth (Pickett and McDonnell, 1993), urbanisation of the landscape and significant losses in rural habitat (Cole, 1983). The information produced by such research is also becoming important in urban planning and the effects of humans on ecosystems have become incorporated in of the study of urban ecosystems (Niemela, 1999).

The negative perception of urban habitats continues to be exacerbated by the terminology used and assumptions made. Agricultural grassland with intensive direct human management, often with low biodiversity including non-native species, are referred to as 'greenfield' sites. Previously developed sites which have been left as derelict and have been naturally colonised are referred to as 'brownfield sites'. In the most recent urban White Paper, the UK government argues that 60% or more of new housing must be located on brownfield sites, an approach predicated on the view that the former are more valuable than the latter. But is this necessarily so?

Many conservationists and ecologists appear to have adopted the same biased view, as evidenced by the number of papers written about the ecology and conservation of agricultural systems compared to urban systems. Since Tansley (1939), habitats have been described as natural, semi-natural and artificial, the value of sites being linked to the degree of human interference and modification (e.g. Ratcliffe, 1977), with natural habitats having the greatest value and artificial the least. The latter category is, however, often used to describe urban and ex-industrial habitats (e.g. Ratcliffe, 1977) even if they are a product of natural and colonisation and have been left undisturbed or managed since dereliction. Yet a regularly managed arable habitat can be described as being semi-natural. The use of term artificial brings with it negative connotations and assumptions which, in the case of urban habitats, perpetuates their second-class status. Is such differentiation based on previous, rather than current, land use and human impacts, appropriate?

Adams (2003) notes that formal conservation is concerned with 'nature' outside the urban environment. Our national conservation priorities (Ratcliffe, 1977; JNCC, 1989) reflect a clear rural and semi-natural habitat focus, and such proprieties are in turn reflected in formal conservation designations, but as can be seen later, some urban habitats have been afforded designation. Urban sites may not be 'natural', but if what Ratcliffe (1984) argues is true, i.e. that there may be no extant 'natural' sites in the UK, there is no justification for considering a naturally colonised urban wasteland (brownfield site) to be less valuable than an intensive agricultural greenfield site.

## Urban ecological approaches

In the past 'ecologists have primarily sought to understand their subject of study in the absence of humans and have considered humans chiefly as agents of disturbance' (Pickett and McDonnell, 1993). Such a desire to investigate 'pristine' environmental systems would appear to preclude the study of urban ecology. Urban ecology research has, however, been undertaken for over fifty years. Early studies into urban ecology were largely descriptive (McIntyre, Knowles-Yanez and Hope, 2000). By the 1980s, however, authors such as Gilbert (1989) sought to synthesise urban ecological knowledge, often using a landscape/ land-use approach. Gilbert's (1989) classic *The Ecology of Urban Habitats* for the first time provided the same basis for the ecological understanding and study of urban habitats as that provided for rural habitats by Tansley in 1939.

Recent urban ecological studies have been more quantitative and have been established within an 'urban ecology framework' (McIntyre, Knowles-Yanez and Hope, 2000). Authors have also applied existing theories such as 'island biogeography' and meta-population theory (e.g. Hanski and Simberloff, 1997) to explain urban ecosystems. Others have taken a more biogeographical approach e.g. Weighmann's (1982) and Schaefer's (1982) work on species richness and habitat size, and Dawson's (1994) work on the impact of 'corridors and boundaries' in relation to reducing species isolation in cities. Niemela (1999) argues that the 'successful study of urban ecosystems' can be achieved using 'existing ecological theories' but that the 'intense human presence' in urban ecosystems requires the development of a 'human ecosystem model', which emphasises the 'social components with connections to ecology'. Yet expanding ecological theory to incorporate urban settings remains new territory for advanced research agendas.

Rebele (1994) categorised urban ecological research into two broad orientations: ecology and social sciences, and this can now be expanded to include planning and interdisciplinary approaches. Indeed some authors argue that the proper study of urban ecology demands such an integrated approach (e.g. Niemela, 1999). This synthesis of social and biological orientated ecology in considering urban areas could be utilised in local urban-based community initiatives linked to regeneration (e.g. Steel, 1999) and Agenda 21.

An important branch of research into urban conservation considers the social implications of open spaces and nature in urban areas. As early as 1890, Olmsted recognised that 'vegetation in cities has a social, medical and psychological role'. A study by Burgess et al. (1987) in south London, for example, focused on the 'meanings and values' people afford open space. The research identified three 'popular values' urban dwellers associated with nature sites: 'wildlife is fun', a 'desire for adventure, and a 'search for variety'. The last twenty years have also seen the development of urban ecological and conservation survey methods which recognise the distinct characteristics of these environments (e.g. Freeman, 1999). These include:

- West Midlands County Council Survey 1982–84
- Leicester Habitat Survey 1983
- Greater London Habitat Survey 1984/85

- Sheffield Inner Habitat Survey 1989
- The Leeds Ecological Advisory Service – Freeman 1999
- Rohde & Kendal – English Nature Report No 22
- Comedia *The Future of Urban Parks and Open Spaces in the UK* (2000)

The coverage of some UK urban ecological surveys is summarised in Table 1, overleaf.

As indicated in Table 1, some recent urban surveys look at both ecological aspects such as habitat type and social factors such as use and attitudes (e.g. Greater London Authority 2001; Venn, 2001).

## The value of urban ecology

Assumptions about the urban environment and urban habitats, as noted earlier, generally fall into two opposing camps. In fact, it would be quite easy to assume that these camps are talking about two entirely different habitats. The common myth suggests that urban landscapes are concrete jungles, paved wall-to-wall with tarmac – a hostile environment where nature struggles to survive, yet Baines (1999) notes that even the most densely built-up places have abundant open space and, in a typical western town or city, green space is the majority land cover.

Increasingly, surveys and research are demonstrating the validity of the latter viewpoint. Despite a number of well-known problems associated with urban areas e.g. pollution, waste, poor soil quality caused by compaction and impacts on hydrology, urban areas support a diversity of species in areas of semi-natural habitat, wasteland, park and other highly human-influenced biotopes (Niemela, 1999). Shepherd (1994) recognised that the diversity of species in urban landscapes 'can often be high, and include rare and threatened species', a view which is supported by past and current research. Gemmell (1977), for example, noted the value of industrial habitats 'for the conservation of uncommon species'. Eighty percent of Bee Orchid sites in the West Midlands are on ex-industrial land. Eversham *et al*. (1996)

# Table 1: Characteristics, methodology and date coverage of some UK urban surveys

| Site Survey Name | Characteristics | Methodology | Data Gathered |
|---|---|---|---|
| Birmingham–West Midlands (1982–84 and 1987–88) | An ecological survey with a habitat classification system adapted for the West Midlands area | 20,000 sites (over 0.5 ha) surveyed. Aerial photographs used to supplement site visits. Repeated in 1985 & 1988 excluded amenity areas, improved grassland and other areas not considered to be important for nature conservation, all sites visited in this case. | Ownership, Planning History, Size Topography Reference Data, Land use, past and present Habitat Classification Species Observations, Signs of community use, types of use, observed threats pollution, litter. |
| Leicester Inner City Survey (1983) | An ecological survey of the Leicester city area | Ariel photographs supplemented by visits to 1,800 sites. Areas coded into habitat categories. Visit to detail plant species | Habitat and plant species |
| The Greater London Habitat Survey (1984/85) | An ecological survey to audit wildlife and habitat information | Phase One methodology with adaptations to suit urban environments. | Land use and Planning status, Ariel Photograph Habitat, species (plus invasive species) Public Access, Amenity Use, Level of Use. Ecological Threats |
| Leeds (Freeman) – 1999 | A site survey and evaluation method developed to provide planners with the information necessary to evaluate the ecological value of sites proposed for development. | 30 sites visited with naturally regenerating habitat and scheduled for development. A site record, evaluation matrix and set of associated values was developed for each site. These values were then used to identify and compare sites. | Age, Size, Neighbour and Use, Land use, Habitat Classification Species, Contribution of urban species Access to site, Site amenity use (19 classes), Level of Use Legal designation, Site Management activities Observed Threats. |
| Sheffield Inner City Survey | An inner city survey of sites for ecological data | Areas selected within the city for site visits | Species, Habitat, Land use, and Threats |
| Greenwich – South London Burgess (1989) | Survey to identify the perceptions and values held by people in relation to areas of open space | Focus Group Discussions and interviews | Description of perceptions and feelings in connection with open space experiences. |
| Redditch Cole and Bussey – The value of urban woodland (1993) | A survey of the social attitudes towards urban woodland, identifying the key characteristics important in making woodland of high social value. | 592 questionnaires with qualitative information on user emotions taken from 295 interviews from residents, near 18 selected woodlands. Interview quotations serve to identify the nature of the woodland experience in terms of individual personal valuations. | Structure of woodland estates. Major species Emotional and cultural responses to woodland. Amenity use of woodland. |
| COMEDIA - The 'Public Parks Assessment' (2000) | Nationally Based survey of metropolitan parks | The study obtained information about the park services provided by over 400 of the larger local authorities | Land Type/Finance/ Designations Assessment of stock Number of Visitors Facilities in park Historic Importance Evidence of a park strategy. |

reported that man-made habitats, such as roadside spoil heaps and quarries could support up to 35% of the rare carabid beetle species in Britain. The importance of wasteland sites for invertebrates has been illustrated by a number of studies, e.g. Gibson (1998) identified that 12%–15% of all red data book invertebrates occur these sites. Urban habitats can support remnants of native flora and fauna (London Biodiversity Partnership, 2001), urban 'specialists' (LWT, 2000), and non-natives (Gilbert, 1989). The London Biodiversity Audit (London Biodiversity Partnership, 2001) recognises that 'urban wasteland ... may be ... one of the most diverse of London's habitats, important for ruderal or pioneer species and invertebrates'. As stated by Adams (2003), 'nature is also there in our cities, [it is] as vital, and sometimes even more powerfully, than outside' (Adams, 2003).

Despite this evidence, even within urban areas the highest conservation value is often bestowed on 'encapsulated countryside', i.e. those parts of the rural environment which have survived urbanisation as remnant habitat islands. Although 'artificial' urban habitats such as derelict land are frequently viewed as being ugly or unattractive (e.g. Civic Trust, 1964), they also provide opportunities for a variety of species to establish in urban areas, creating their own natural 'wilderness' (e.g. Teagle, 1978; Goode, 1986; Gilbert, 1989; Morphet, 1990).

Back in the 1960s McHarg (1969) recognised that we need nature as much in the city as in the countryside, a sentiment that continues to be repeated, e.g. 'People in towns and cities have a special need for contact with nature and urban green space' (Countryside Commission, 1990). The social value of urban ecology for amenity, education and scientific purposes has also been well reported (e.g. Davis, 1976; Tregay, 1986). Again, these values continue to be reiterated, and research continues to demonstrate the physical and psychological benefits of contact with urban nature and open space. Researchers (e.g. Mostyn, 1979) have determined that both

formal open spaces and informal derelict land can have a conservation value to local communities. Urban nature provides colour and variety in rather monotone built-up areas and is recognised as making a significant contribution to the quality of urban life (e.g. Kaplan, 1984; Barker, 1989; Harrison, 1993; Huang *et al.*, 1992; Rohde and Kendle, 1994).

At least 4% of all biological SSSIs are on 'damaged land' (CORDATA), most within an urban context, yet conservation criteria (e.g. Ratcliffe, 1977; JNCC, 1989) are biased towards traditional rural habitats. Such nature conservation values are determined by a range of factors including 'aesthetic, nationalistic and cultural values as well as purely ecological ones' (Toogood, 1997) and reflect an underlying, often negative, attitude towards the urban environment.

## Case study in urban ecological and conservation value

Greater London provides an illustration of the value of the conservation and ecology resource that is present in our cities. The conservation designations applied to Greater London include: a Special Protection Area under the Birds Directive, (SPA), two Special Areas of Conservation under the Habitats Directive (SAC), two National Nature Reserves (NNR) and 33 Sites of Special Scientific Interest (SSSI). In addition there are over 1200 Sites of Importance for Nature Conservation covering an area of 28,000 hectares.

The 1984/85 the Greater London habitat audit identified 15–20 separate habitat types in London (LWT, 1984), from ancient woodlands, to inner city 'wasteland'. The ecologically important habitats found in London, as identified by the London Biodiversity Partnership (2002) for the preparation of the London Biodiversity Audit, are listed in Table 2.

In 2000, the Greater London Assembly estimated that, in London's 33 boroughs, nearly 158,000 hectares (over 600 square miles and almost 17%) of this land offered opportunities for environmentally based schemes. The London Biodiversity Partnership (2001) also stated that 'up to 40% of Greater London can be designated as green open space', with almost half of this having some value as wildlife habitat. Over 1500 species of flowering plants and 300 types of bird are found in the capital (GLA, 2000).

## Conclusion

The number of people living in urban areas in Britain is increasing; the amount of urban land cover is also increasing. The National Office for Statistics (1999) predicts that urban land cover in the United Kingdom will increase from 14.9% by at least a further 20% by mid-21st century. Developing our knowledge of the ecology and ecological processes associated with densely populated areas is essential if we are to understand and manage our environments.

Historically, Fitter (1945) admonished ecologists who lived in urban areas for their 'shameful neglect [of the] flora of the waste sites that lie under their noses'. Freethy in 1986, noted that 'in summer many urban-based naturalists head for the surrounding countryside and leave what they describe as a lifeless town behind them'. Is the situation any different today? It can be argued that if urban conservation and ecology are to develop then the negative attitudes of many planners, ecologists and conservationists will need to change.

The UK Biodiversity Steering Group Report (DoE, 1994) provides a list of thirty-seven habitats for which conservation action is required. Thirty-seventh on the list is urban habitats, and their inclusion represents a significant, and positive, step regarding the recognition of their existence and value. It is to be hoped that more urban ecological researchers and conservation practitioners, will not only recognise the importance of the urban environment and its habitats, but will act to ensure that the immense values bestowed in terms of nature conservation and quality of life are incorporated in urban design and development.

**Table 2: Habitat and Land Use Classification for London (London Biodiversity Partnership 2002)**

| Habitat Type | Description |
|---|---|
| Woodland | Woodland and scrub habitats: ancient, secondary, 'recent', wet etc. |
| Open Landscape with Ancient/Old Trees | Including deer parks, old parkland, wood pasture, other areas with unimproved grassland and scattered old/ ancient trees |
| Hedgerows | Various boundary features with trees and shrubs |
| Acid Grassland | Unimproved and semi-improved grassland on nutrient-poor, free-draining soils (e.g. sands and gravels) |
| Chalk Grassland | Unimproved and semi-improved grassland on chalk |
| Grassland. Meadows & Pasture | Unimproved and semi-improved grassland other than acid grassland, chalk grassland or wet grassland, i.e. neutral grassland |
| Heathland | Sites where heather occurs naturally |
| Grazing Marsh & Floodplain Grassland | Sites where the habitat is dependent upon a combination of periodic wetting or inundation and grazing or cutting |
| Marshland | All wet terrestrial habitats including, fens, bogs, mires, swamps |
| Reed Bed | Sites where common reed is dominant. |
| Rivers & Streams | All free-flowing watercourses above the tidal limit |
| The Tidal Thames | All areas of the river Thames and its tributaries below the tidal limit |
| Canals | The London canal network |
| Ponds, Lakes & Reservoirs | All standing open water |
| Private Gardens | |
| Parks, Amenity Grassland & City Squares | All formally managed amenity open space (including sports pitches, school grounds and landscaped areas around institutional buildings) |
| Railway Lines & Sidings | All vegetated or natural surfaces within rail side boundary fencing |
| Urban Wastelands | Semi-natural vegetation which has developed on an imported or artificial substrate resulting from previous development or disturbance |
| Farmland | Arable fields and agricultural lees |

# References

Adams, W.M. (2003) *Future Nature: a vision for conservation* (Volume 2). Earthscan, London.

Adams, W.M. (1996) *Future nature: a vision for conservation*. Earthscan, London.

Baines, C. (1999) *Urban Areas*. In: Sutherland, W. J. (ed.) *Managing Habitats for Conservation*. CUP, Cambridge.

Baines, C. (1986) *The Wild Side of Town*. BBC, London.

Barker, G. (1989) People and wildlife in cities. Women's Coruna Society 9th Triennial Conference, Westminster, 13.7.89 (published informally in notes of conference).

Botkin, D.B. (1990) *Discordant Harmonies*. Oxford University Press, New York.

Burgess, J., Harrison, C.M. and Limb, M. (1988) People, Parks and the Urban Green: A study of Popular Meanings and Values for Open Spaces in *City Urban Studies*, **25**, 455–473

Cairns, J. (1987) *Disturbed ecosystems as opportunities for research into restoration ecology*. In: Jordan, W.R., Gilpin, M.E. and Aber, J.D. (eds) *Restoration ecology: a synthetic approach to ecological research*. Cambridge University Press, Cambridge.

Carr, A. and Lane, S. (1993) *Practical conservation: urban habitats*. Hodder and Stoughton, London.

Civic Trust (1964) *Derelict Land*. Civic Trust, London

Cole, L. (1983) *Urban Nature Conservation*. In: Warren, A. and Goldsmith, F.B. (eds) *Conservation in Perspective*. John Wiley and Sons Ltd

Greenhaigh, L. & Walpole, K. (1995) *Park Life: Urban Parks and Social Renewal*. Comedia with Demos, Stroud and London.

Countryside Commission (1990) *Countryside and nature conservation issues in district local plans*. Countryside Commission, Cheltenham

Davis, B.N.K. (1976) Wildlife urbanisation and industry, *Biological Conservation*, **10**, 249–291

Dawson, D. (1994) *Are habitat corridors conduits for animals and plants in a fragmented landscape?* English Nature Report 94, EN, Peterborough

Department of the Environment (1994) *Planning and Policy Guidance: Nature Conservation* (PPG 9) HMSO, London

Douglas, I. (1983) *The urban environment*. Arnold, London.

Emery, M. (1986) *Promoting Nature in Cities and Towns*. Croon Helm, London.

Evans, D. (2003) *A History of Nature Conservation in Britain* (Second Edition). Routledge, London.

Eversham, B.C., Roy, D.B. and Tefler, M.G. (1996) Urban industrial and other manmade sites as analogues of natural habitats. *Carabidae. Ann. Zoo. Fennici*, **33**, 149–156.

Fitter, R. (1945) *London's Natural History*. Collins New Naturalist, London.

Flint, R. (1985) *Encouraging Wildlife in Urban Parks*. The London Wildlife Trust, London.

Freeman, C. (1999) Development of a simple method for site survey and assessment in urban areas. *Landscape and Urban Planning*, **44** (1), 1–11.

Freethy, R. (1986) *Guide to Wildlife in Towns*. British Naturalists Association with Crowood Press, Marlborough.

Gemmell, R.P. (1977) *Colonisation of Industrial Wasteland*. Arnold, London.

Gibson, C.W.D. (1998) Brownfield red data: the values of artificial habitats for uncommon invertebrates. *English Nature Research Report* 273. English Nature, Peterborough.

Gilbert, O. (1989) *The Ecology of Urban Habitats.* Chapman and Hall, London.

Greater London Authority (2000) *Connecting with London's Nature, the Mayor's Biodiversity Strategy.* GLA, London.

Good, R. (2000) The value of gardening for wildlife – What contribution does it make to conservation. *British Wildlife,* **12** (2), 77–84.

Goode, D. (1986) *Wild in London.* Michael Joseph, London.

Hanski, I, & Simberloff, D. (1997). The metapopulation approach, its history, conceptual domain and application to conservation. In Hanski, I . & Gilpin (eds) *Metapopulation Biology, Ecology, Genetics and Evolution.* Academic Press, London.

Harrison, C. (1993) *Nature Conservation, Science and Popular Values.* In: Goldsmith, F.B. & Warren, A. (eds) *Conservation in Progress.* Wiley, Chichester

Howard, E. (1898) *Tomorrow.* Swann Sonnenschein, London

Huang, J. Ritschard, R., Sampson, N. & Taha, H. (1992) *The Benefits of Urban Trees.* in Akbari, H, Davis, S. Dorsano, S., Huang, J & Winnett, S. (eds) *Cooling our cities.* US EPA Office of Policy Analysis, Washington DC.

Joint Nature Conservation Committee (1989) *Guidelines for the selection of biological SSSI's.* JNCC, Peterborough.

Welsh, J. (1989) *Wildlife on the doorstep.* English Nature, Newbury.

Kaplan, R. (1984) *The role of nature in the urban context.* In: Altman, I. & Wohlwill, J.F. (eds) *Human behaviour and environment.* Plenum, New York.

King, A. & Clifford, S. (1985) *Holding Your Ground.* Wildwood House, Hants.

London Biodiversity Partnership (2001) *London Biodiversity Action Plan, Part 1.* London Biodiversity Partnership, London.

London Biodiversity Partnership (2002) *London Biodiversity Action Plan, Part 2 – the action.* London Biodiversity Partnership, London.

London Wildlife Trust (1984) *Greater London wildlife audit.* LWT, London

Mabey, R. (1973) *The Unofficial Countryside.* Collins, London.

Marren, P. (2002) *Nature Conservation.* Collins New Naturalist, London

McDonnell, M.J. (1997) A paradigm shift. *Urban Ecosystems.* **1**(2), 85–86.

McIntyre, N.E. Knowles-Yanez, K. & Hope, D. (2000) Urban ecology in the use of 'urban' between the social and nature sciences. *Urban Ecosystems*, **4**, 5–24.

Morphet, J. (1990) Can we save urban open space. *ECOS*, **11**(4), 27–43.

Mostyn, B. (1979) *Personal benefits and satisfactions derived from participation in urban wildlife projects.* Nature Conservancy Council, Peterborough.

Mumford, L. (1938) *The Culture of Cities.* Seeker and Warburg, London.

Murphy, G. (2002) *Founders of the National Trust.* (Second Edition). National Trust, London.

Nature Conservancy Council (1984) *Nature Conservation in Britain.* NCC, Peterborough.

Nicolson-Lord, D. (1987) *The Greening of Cities.* Routledge, London.

Niemela, J. (1999) Is there a need for a theory of urban ecology? *Urban Ecosystems*, **3**, 57–65.

Pickett, S.T.A. & McDonnell, M.R. (1993) *Humans as components of ecosystems: The Ecology of Subtle Human effects and Populated Areas*. Springer-Verlag New York, USA.

Porter, R. (1996) *London, A Social History*. Penguin, London.

Ratcliffe, D. (1984) Post-medieval and recent changes in the British vegetation. *New Phytologist*, **98**, 73–100.

Ratcliffe, D. (1977) *Nature Conservation Review Volume 1*. CUP, Cambridge.

Rebele, F. (1994) Urban Ecology and special features of urban ecosystems. *Global Ecology Biogeographical Letters*, **4**, 173–187.

Rees, W.E. (1997) Urban Ecosystems: the human dimension. *Urban Ecosystems*, **1**, 63–75.

Rhode, C.L.A. & Kendle, A.D. (1994) *Human well-being, natural landscapes and wildlife in urban areas. A review. No. 22*, English Nature, Peterborough.

Rodwell, J.S. (ed) 1991 to 2000) *British Plant Communities*. Volumes 1 to 5, CUP, Cambridge.

Schaefer, M. (1982) Studies on the arthropod fauna of green urban ecosystems. In: *Urban Ecology*, The Second European Ecological Symposium, Blackwell, 65–73.

Sheil, J. (1976) *Nature in Trust: The History of Nature Conservation in Britain*. Brocke, Glasgow.

Shepherd, P.A. (1994) A review of plant communities of derelict land in the city of Nottingham, England and their value for nature conservation. *Memorabilia Zoological*, **49**, 129–137.

Smyth, B. (1987) *City Wildscape*. H Shipman, London

# Urban history and the evolution of townscapes

## Melvyn Jones
Sheffield Hallam University

## Abstract

Urban areas all over the world can be seen to have gone through a number of phases of economic and social development which has resulted in enormous spatial expansion and internal change. This paper focuses on the British city and aims to show the phases of growth and change through which they have passed in the last 200 years and the typical townscape elements that have developed as a result of this evolution (Hornby & Jones, 1991). In view of the context in which the paper is given, particular attention is given to the green environment and historical ecology.

## Early towns

Most of our towns and cities have a medieval origin and grew up around a church and market place often under the protection of a castle. Beyond this central focus lay a maze of streets, particular ones often dominated by a specific craft or trade. And beyond may be a town wall. Variations on this theme of location on an important route (for example at the terminus or meeting of routes or at an important water crossing) with church and market dominating the town plan and both often under the protection of a castle are found throughout Europe and elements of this general type can even be detected in the earliest towns founded in North America.

Such compact and overwhelmingly 'pedestrian' towns began to change in the eighteenth or more commonly in the nineteenth century (Figure 1a). By 1800 the densely developed urban area was surrounded by a fringe-belt in which were located a number of urban-oriented land uses which developed there because they had been excluded from the town, for example, (isolation hospitals or noxious industry) or because they preferred to distance themselves from the urban core (for example, the country estates of rich merchants) or because they required certain resources not available in the tightly packed urban core (for example water power sites for certain industries or farmland in the case of market gardeners or dairy farmers). Figure 2 shows part of the fringe belt of Sheffield in 1832.

## Nineteenth and early twentieth century change

By the mid-nineteenth century important changes were beginning to take place (Figure 1b). Industry was now largely reliant upon steam power and locations near to coal supply depots became important and therefore the beginnings of important industrial zones developed around the old core. In close proximity to the factories, tightly-packed residential areas for the workers were erected mainly in the form of brick terraces. At the same time, the upper and middle classes moved to the former fringe belt on the periphery where large villas were springing up. The former old town was now less of a mixture of industry, shops, offices and residences, and was evolving as the commercial heart of the urban area in which specialist shops, offices and financial institutions were beginning to predominate.

By the beginning of the first decade of the twentieth century the zones that were beginning to emerge 60 years earlier were firmly established (Figure 1c). Housing and environmental standards had risen, however, as evidenced by the provision of public open spaces (parks and purchased or gifted woodlands) and housing that complied with certain building and environmental standards. The rapid outward expansion of the urban area had been facilitated by the development of public transport and this had resulted in the engulfment of farms, hamlets and villages,

# Figure 1. A British city, 1800–2003: a generalised sequence of morphological change.

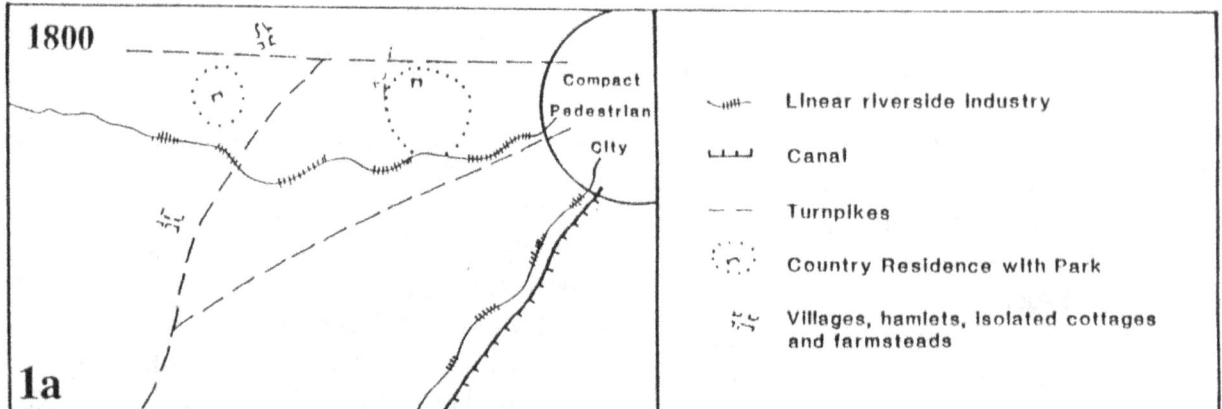

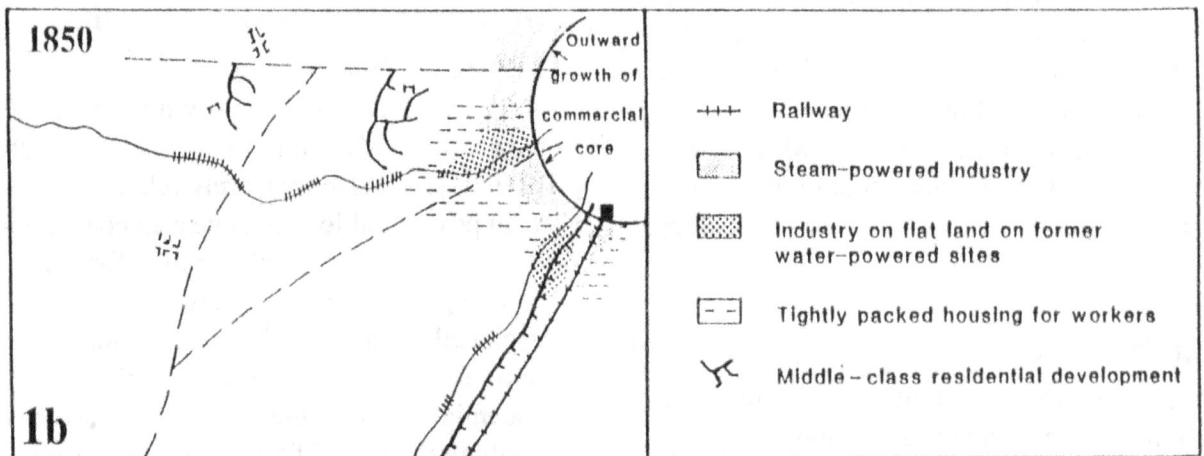

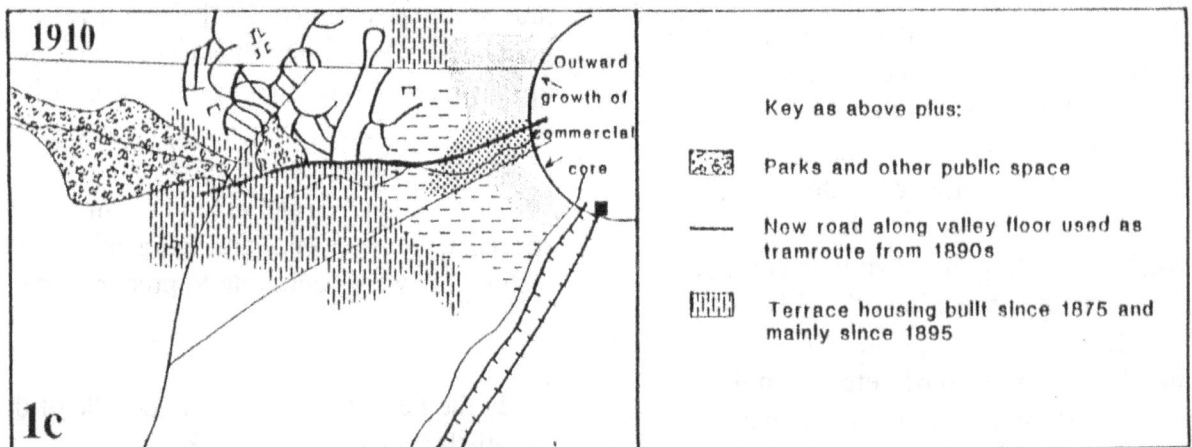

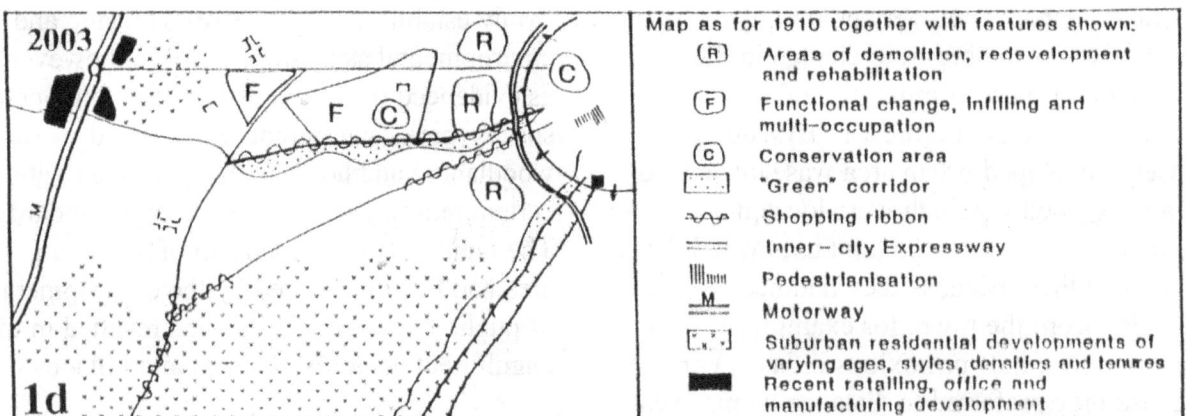

**1a** — 1800

Compact Pedestrian City

- ~ IIII ~   Linear riverside industry
- ⊔⊔⊔   Canal
- – –   Turnpikes
- (⌂)   Country Residence with Park
-   Villages, hamlets, isolated cottages and farmsteads

**1b** — 1850

Outward growth of commercial core

- +++   Railway
- ▫   Steam-powered industry
- ▨   Industry on flat land on former water-powered sites
- [– –]   Tightly packed housing for workers
- Ƴ   Middle–class residential development

**1c** — 1910

Outward growth of commercial core

Key as above plus:

- ▨   Parks and other public space
- ──   New road along valley floor used as tramroute from 1890s
- ▥   Terrace housing built since 1875 and mainly since 1895

**1d** — 2003

Map as for 1910 together with features shown:

- (R)   Areas of demolition, redevelopment and rehabilitation
- (F)   Functional change, infilling and multi-occupation
- (C)   Conservation area
- [∴]   "Green" corridor
- ∿∿   Shopping ribbon
- ──   Inner–city Expressway
- ▥   Pedestrianisation
- M   Motorway
- [∴]   Suburban residential developments of varying ages, styles, densities and tenures
- ■   Recent retailing, office and manufacturing development

and to the emergence of a star-shaped city, as urban development has followed the radial rail and tram routes into the countryside.

## The evolution of the modern urban area

Between 1910 and the present day the urban area, mainly in the form of residential suburbs has continued to spread outwards, at first mainly in a ribbon form along the main transport routes but later more compactly as green belt legislation has prevented further expansion (Figure 1d). On the main roads leading from the centre, retailing ribbons, extend well into the suburbs occasionally swelling to become substantial shopping centres. More parks and woods are present within the urban landscape together with arcas of entrapped countryside

The more central parts of the city have also undergone dramatic changes. Pedestrianisation of some parts of the central shopping area has taken place and this, together with the inner-city expressway, helps to ease traffic congestion.

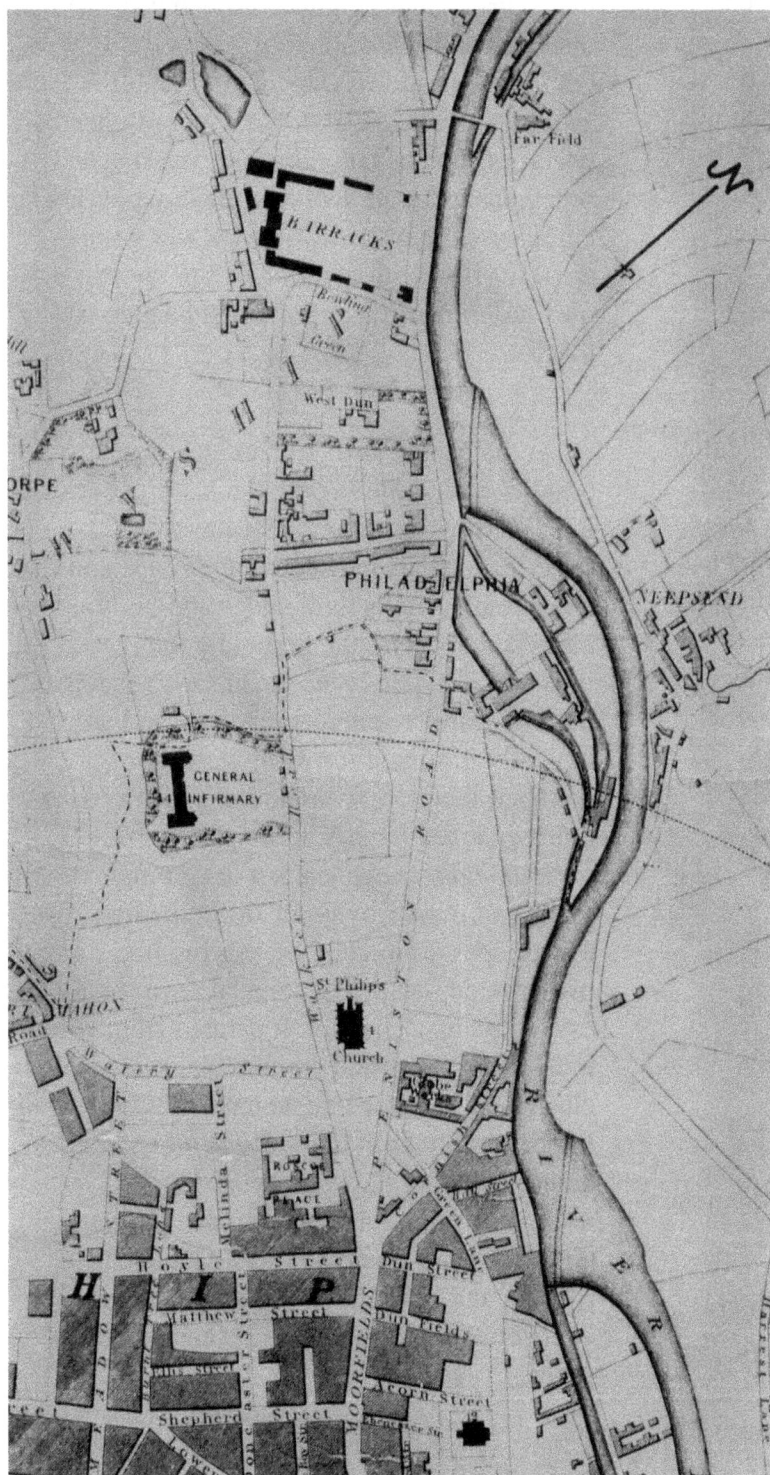

**Figure 2. Part of Sheffield's fringe belt in 1832 as portrayed on J. Tayler's map of the town and its environs. It contains a water-powered industrial site, the General Infirmary and the military barracks.**

Another characteristic of the central area by the beginning of the twenty-first century is the concentration there of high-rise office and public buildings. In that part of the inner city outside the central commercial core lies a zone of decay, change and improvement. In some parts old workshops and warehouses are giving way to shops and offices as the commercial core expands, in others early terraces have been replaced by high-rise residential developments or new factory units. In some areas housing improvement schemes have given a new lease of life to old residential areas. Some of these have been 'gentrified', as a result of which young middle-class residential households have gradually displaced working-class households. Some of the larger Victorian villas have undergone changes of function and are now offices, or have been converted into bed-sitters.

More dramatically, in visual terms, former industrial premises such as mills and warehouses have been converted to residential use, especially where they overlook water along canals and rivers. Advantage has also been taken of the decline and abandonment of industrial properties along rivers and canals to 'green' inner city environments in order to create more pleasant surroundings for inner-city residents and workers and to make such areas more attractive to developers and manufacturers. Such projects have involved cleaning up rivers and canals, stabilising banks, landscaping, seeding and planting, and creating riverside walks.

Perhaps the most dramatic morphological changes in recent years have taken place in the urban periphery, especially where motorways approach the edge of the urban area. Not only have such areas seen a dramatic expansion of residential development to suit the needs of young house buyers requiring cheap first homes near to good communications networks, they have also provided attractive sites for industry, offices, warehouses and distribution depots and for new retail developments. Large superstores are now commonplace and out-of-town

'regional centres' that rival town-centre shopping areas in their size and range, and with massive free car-parking facilities have become major landmarks.

## The development of townscapes

One of the best-known attempts to generalise about townscape was made by Professor A.E. Smailes nearly 50 years ago (1955) and his scheme is still relevant today. Smailes' research led him to view the townscape of a British town or city as a series of imperfect concentric zones (Figure 3). Each zone will be distorted and interrupted by major communication links, by natural features such as rivers and lakes, and by the persistence of enclaves such as commons, university precincts and engulfed villages. Smailes also emphasised that the townscape will be in constant flux with continuous but uneven outward expansion and internal change taking place.

At the centre of the urban area is what Smailes called the *kernel*, the historic core, identified by its commercial function, its street pattern, its individual buildings, and the mixture of architectural styles and building materials. Beyond the kernel lie the integuments. The *inner integuments* are mainly of nineteenth century date and are typically of mixed use with workshops, cramped factories, warehouses, railway land, tightly packed houses and high flats. The *outer integuments* present a less complex picture of more obviously differentiated townscapes dominated by varying tracts of owner-occupied and municipal residential estate layouts. They are separated from industrial estates and commercial ribbons and clusters. The outer integuments merge imperceptibly with the *urban fringe*, Smailes' outermost zone in which surviving farmhouses and farmland, market gardens and allotments are found amongst residential ribbons and outliers, urban utilities such as cemeteries and sewage works and amenity space.

**Figure 3. Components and characteristics of British townscapes.** *Source*: **simplified from Smailes.**

|  | | Growth phase | Functional zone | Relief | Fabric materials |
|---|---|---|---|---|---|
| **K E R N E L** | | Old town | Enclave(s).<br><br>Commercial *core* with prongs outside kernel. Ousting of residence. | Architectural dominants, e.g. castle towers, church spires.<br><br>High buildings: point blocks; rugged profile; irregular street cornice. | Traditional, or imported stone.<br><br>Concrete, replacing traditional materials. |
| **I N T E G U M E N T S** | | Late 18th and early 19th century | Decayed inner zone of mixed use. Workshops, cramped factories, warehouses, high-density residence (slums and high flats); railway space.<br><br>Professional and administrative quarters. | Low buildings; little relief except churches; terrace housing in formal layouts; some replacement with flat blocks. | Mixed.<br><br><br>Brick or stucco and slate. |
| | | Railway age before 1914 | Industry and tightly-packed housing mixed. | Terrace-ribbing with factory and neo-Gothic church salients.<br><br>Gas-holders. | |
| | | Railway and automobile age since 1919 | Industry and housing segregated.<br><br>Villa housing in open, bourgeois suburbs and municipal estates.<br><br>Spacious factory lay-outs<br><br><br>Village enclaves. | Villa-studding.<br><br>Intermixture of roofs and tree tops.<br><br>Factory scaling: low buildings with extensive continuous roof surfaces.<br><br>Power-station chimneys and cooling towers. | Brick, rough cast, and tile.<br><br>Much foliage.<br><br><br>Traditional. |
| Urban fringe | | | Interim development; residential ribbons and outliers; urban utilities, e.g. cemeteries, sewage works; amenity space and surviving farmland, allotments and market gardens. | | |

*Note* In Scotland, residence in inner integuments is represented mainly by high tenements, exclusively in stone. Tenements continue as important building forms after 1919, and rough-cast brick buildings only recently modify the stone tradition.

Smailes associated distinct townscape forms with the first three of his four zones. In the kernel he drew attention to dominant historical features such as castle keeps, church spires and high Victorian and Edwardian blocks with rugged profiles and irregular cornices, and modern town blocks of concrete and glass. In the inner integuments, row upon row of two-storey terrace housing in formal layouts was in the 1950s in Smailes' view the dominant form, although this monotonous townscape was interrupted by factory chimneys, church spires and towers and the rugged outlines of Victorian schools and chapels. Beyond, in the outer integuments the terrace house is replaced by the villa in the form of detached and semi-detached houses and bungalows in brick and tile set in their own gardens along tree-lined streets. When multiplied over tracts of the urban area the blocks, terraces and villas produce distinctive features of relief which Smailes called *block clumping*, *terrace ribbing* and *villa studding*.

## The green urban environment

Within the modern townscape the green environment survives in a number of historical and modern and semi-natural and artificial forms. They include a remarkably diverse range of historic landscape components including field systems, commons, former deer parks and chases, landscaped parks and historic gardens, ancient woodlands and prehistoric earthworks (for South Yorkshire, for example see Jones, 2000 and 2003). In addition there are extensive areas associated with early industrial activity especially early mining landscapes, water transport arteries and river valleys with water-power sites.

In very general terms it is possible to identify three types of urban countryside. In the *urban fringe countryside*, the working countryside and surviving woodlands are inter-mixed with enlarged villages, outer urban housing estates, industrial estates and recent retailing developments, with urban utilities such as cemeteries and sewage farms and with allotments and country parks. Odd corners of countryside are often cut off by intersecting railways, arterial roads and motorways, but in other parts of the urban fringe large areas of working rural countryside survive, as in the case of the area of countryside shown in Figure 4. This shows an area lying less than five miles (6 km) from the centre of Sheffield in 1894. Apart from infilling and expansion around the villages of Grenoside and Ecclesfield, the ancient landscape of hamlets and scattered farms in local stone, ancient woods that had been coppiced for centuries until the late nineteenth century, an intricate pattern of irregularly-shaped fields, bounded by stone walls and hedges dotted with mature trees, all bound together by a network of winding lanes and footpaths bordered in many places by banks of wild flowers, grasses and ferns, still survives at the beginning of the twenty-first century.

The *suburban countryside* contains greenspace mainly in the form of municipal parks, green corridors along rivers and canals, woodlands, old meadows, pasture and heath, school and college playing fields, and greens and commons associated with villages now engulfed by outward suburban growth – all set within a matrix of residential development that has evolved in the last 75–100 years. An important green element within this sea of housing is the private garden.

As the urban centre is approached the intensity of urban development increases and terrace houses, high-rise blocks and industrial complexes replace the suburban villas and bungalows and, as the commercial heart of the urban area is reached, high buildings and transport arteries dominate the townscape. This is the location of the inner city countryside. It may include parks, but if they exist they are, with some notable exceptions, smaller than their suburban counterparts.

Woodland is usually absent. As part of post-industrial greening projects, planting may have taken place along disused canal corridors and on abandoned industrial sites.

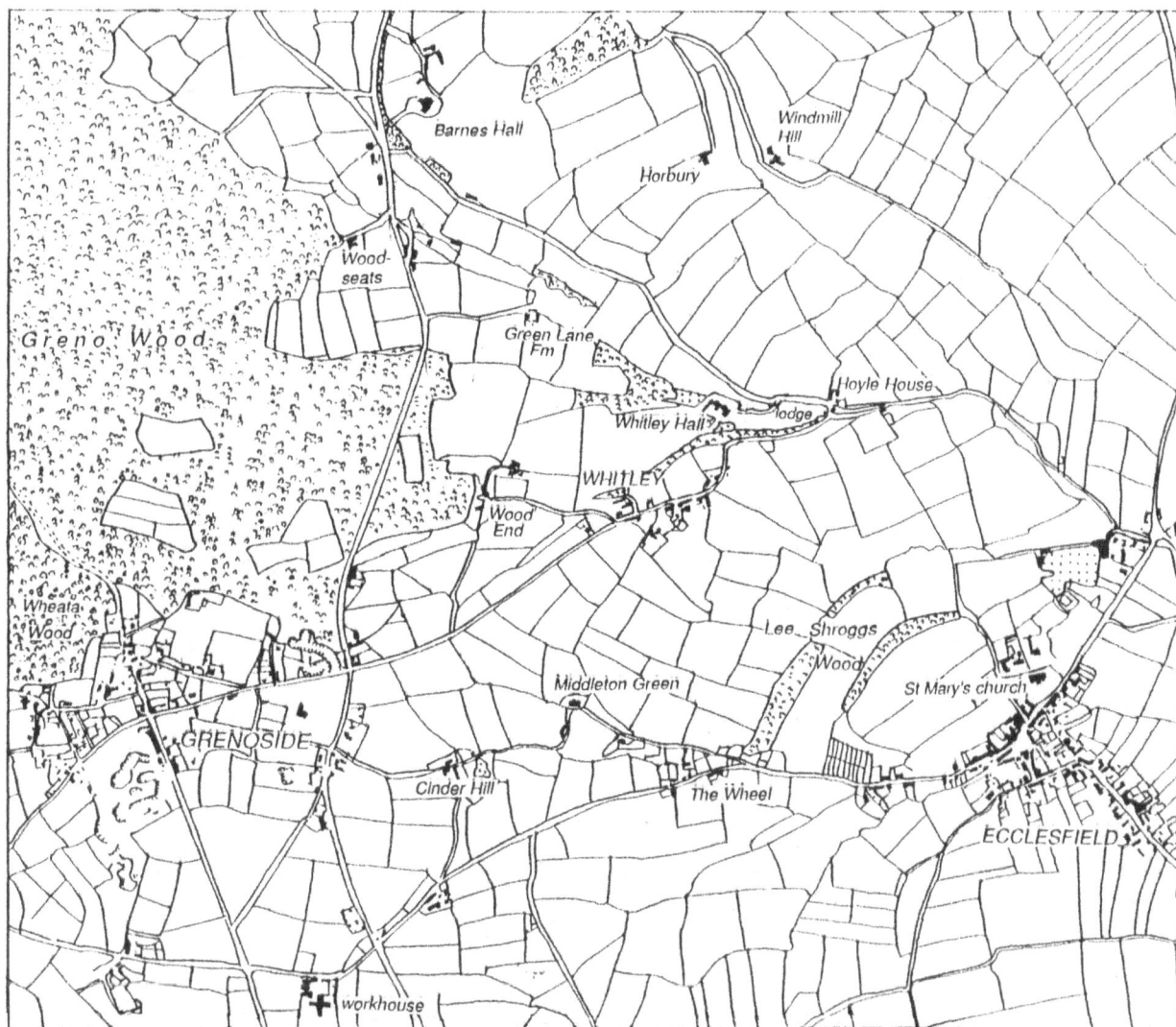

**Figure 4. Part of Sheffield's urban fringe in 1894. The heart of the area is little changed more than a century later. Source: Ordnance Survey Six-Inch map**

Other former industrial sites, awaiting development, contain various kinds of unofficial countryside (called urban commons by Oliver Gilbert (1989; 1996)) from bare ground, through flower-rich grassland, to scrub woodland. These post-industrial landscapes often contain a wetland component in the form of subsidence flashes, ponds and reservoirs.

Post-industrial sites of varying ages, survive not only in the inner city, the suburban areas (more rarely) and in the urban fringe. Figure 5, for example, shows a range of early sites of ironstone and coal mining in the form of spoil heaps from bell pits and shaft mines dating from the early medieval period until the nineteenth century, surviving within a historic landscape that has become suburbanised. Scholes Meadow was created through the felling of the western half of

Scholes Coppice in order to opencast mine the area for coal in the late 1940s. The wetter areas of the meadow now support a profusion of Common Spotted Orchid and Northern Marsh Orchid, the latter being a rare species in Rotherham (Rotherham MBC, 1996). The significant wildlife interest of old mining sites is being increasingly recognised. Lunn & Wild (1995), for example, in their study of thirty South Yorkshire sites, demonstrated that they contained exciting and diverse ecosystems ranging from bare ground to species poor *Deschampsia flexuosa* grassland to scrub dominated by birch and even mature woodland. Sites contained rare and locally scarce plant species such as Bee, Northern and Southern Marsh, Spotted and Pyramidal Orchids; five sites supported Little Ringed Plovers (British population, 600 pairs) and abundant butterfly populations including the scarce Dingy Skipper at two sites. The authors

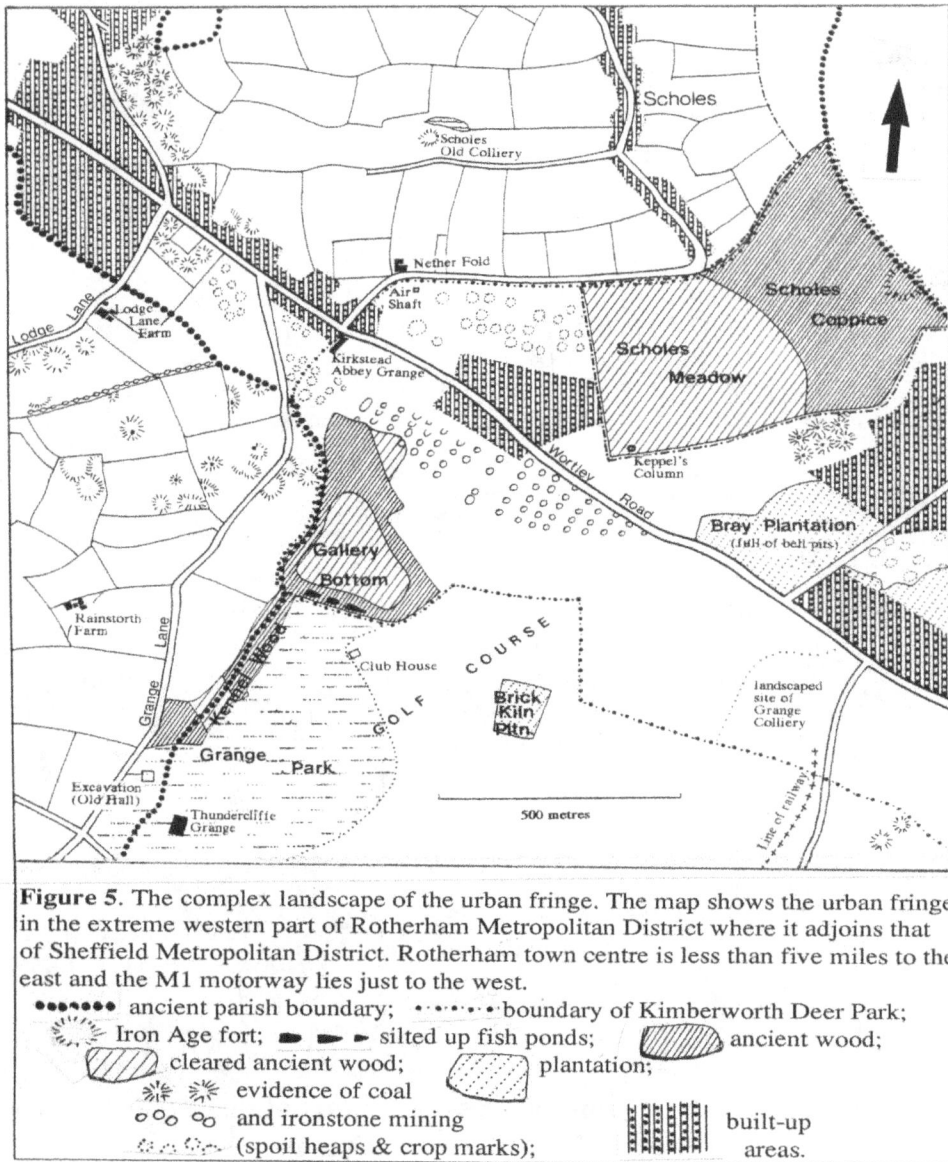

**Figure 5**. The complex landscape of the urban fringe. The map shows the urban fringe in the extreme western part of Rotherham Metropolitan District where it adjoins that of Sheffield Metropolitan District. Rotherham town centre is less than five miles to the east and the M1 motorway lies just to the west.
●●●●●●● ancient parish boundary; •••••••boundary of Kimberworth Deer Park; ⁓Iron Age fort; ➧ ➧ ➧ silted up fish ponds; ⊠ ancient wood; ⊠ cleared ancient wood; ⊡ plantation; ❋ ❋ evidence of coal ⚬⚬⚬ ⚬⚬ and ironstone mining (spoil heaps & crop marks); ▦▦ built-up areas.

## Concluding remarks

What has been attempted here is to put the urban environment into its historical context: to try to generalise about the urbanisation process and to show the evolutionary stages that every sizeable town or city has gone through since at least medieval times and to describe two general models of that evolution. There is a close relationship between the ecology of urban areas and the stages of morphological change through which all urban areas have passed and continue to pass. This is what gives our urban areas their great ecological variety: old deer parks, ancient woods and encapsulated farmed countryside rub shoulders with abandoned – and colonised – industrial sites, railway tracks and canals which themselves are intermingled with thousands of private gardens and public parks and other public open spaces of varying antiquity and modernity. This great variety has resulted from layer upon layer of urban development. Although many early features have been swept away or are buried beneath later developments, many have survived, are taken for granted as part of the modern scene, and sometimes inappropriately managed. The urban landscape is many-layered. And to understand it we need to peel way the layers.

## References

Gilbert, O.L. (1989) *The Ecology of Urban Habitats*. Chapman and Hall, London.

Gilbert, O.L. (1998) Urban Ecological Distinctiveness. In: *Conference Proceedings, Landscapes – Perception, Recognition and Management: reconciling the impossible?*

The Landscape Conservation Forum/Sheffield Hallam University, 2–4 April, 1996. Wildtrack Publishing, *Landscape Archaeology & Ecology*, **3**, 109–110.

Hornby, W.F. & Jones, M. (1991) Urban Morphology. Ch. 5 in: *An Introduction to Settlement Geography*, Cambridge University Press, Cambridge, 58–74.

Jones, M. (2000) *The Making of the South Yorkshire Landscape*. Wharncliffe Publishing, Barnsley.

Jones, M. (2003) *Sheffield's Woodland Heritage* (third edition). Wildtrack Publishing/ Green Tree Publications, Sheffield.

Jones M. & Rotherham I.D. (1998) Eyes have they but they see not: changing priorities and perceptions of countryside in urban areas. In: *Conference Proceedings, Landscapes – Perception, Recognition and Management: reconciling the impossible?* The Landscape Conservation Forum/Sheffield Hallam University, 2–4 April, 1996. Wildtrack Publishing, *Landscape Archaeology & Ecology*, **3**, 19–24.

Lunn, J. & Wild, M. (1995) The Wildlife Interest of Abandoned Collieries and Spoil Heaps in Yorkshire. *Land Contamination & Reclamation*, **3** (2), 135–137.

Rotherham MBC (1996) *Keppel's Field Local Nature Reserve: Draft Summary Management plan 1996–2002*. Rotherham MBC, Rotherham.

Smailes, A.E. (1955) 'Some reflections on the geographical description and analysis of townscapes', *Transactions of the Institute of British Geographers*, **21**, 99–115.

# Issues in urban nature conservation (plants and vegetation)

**Ian C. Trueman**
School of Applied Sciences, University of Wolverhampton

## Abstract

The idea of conserving nature has to be re-evaluated in the urban context. The principal organisms in the urban environment are people, and the study of urban ecology is largely the study of anthropogenic effects. However botanical recording in the urban area readily demonstrates high plant species-richness per unit area, quite comparable with countryside levels (Table 1). Some of this is undoubtedly due to the presence of large numbers of relatively small anthropogenic biotopes, each with its own set of characteristic species. A vast rural forest might include more species than a tiny urban copse, but alongside the latter there might well be canals, lawns, car parks, rubbish dumps, road verges, graveyards and umpteen idiosyncratic gardens all in the equivalent area. Apart from the opportunities given for plant colonisation and survival, the structural complexity must also increase the habitats available for all kinds of organisms.

## Introduction

Modern urban improvement schemes tend to sweep away this complexity and replace it with vast shed cities or neat neo-suburbias. Such is the complexity of city life that the pattern might re-establish itself eventually, but we should try to make landscape architects (and maybe architects and designers of roads) consider how to engineer a little more environmental complexity into their new worlds. As a step in the right direction, the Deposit South Staffordshire Local Plan has included networks of domestic gardens, small natural and semi-natural areas and also road verges and trees as '*features in the landscape of value to wildlife requiring their integrity and continuity to be protected*'.

## Issues and opportunities

Individual species-rich habitats also occur in urban landscapes. There are many urban examples of classical semi-natural vegetation (fondly known as '*trapped countryside*') and they are at least as difficult to conserve as rural examples. The advantages in protecting

**Table 1: The ten most species-rich tetrads in Montgomeryshire**

| Tetrad Grid Reference | No. Mapped Species | Locality | Habitats |
|---|---|---|---|
| SJ22Q | 275 | Llanymynech Hill | Rural (post-industrial) |
| SJ22K | 246 | Llanymynech Hill | Rural (post-industrial) |
| SO29X | 224 | Cornden Hill | Rural |
| SJ31C | 203 | Breidden | Rural |
| SJ20I | 201 | Welshpool | Mostly urban |
| SJ11S | 189 | Near Meiford | Rural |
| SN98M | 198 | Llanidloes | Mostly urban |
| SJ20D | 188 | Near Welshpool | Urban fringe, mostly rural |
| SO08J | 188 | Llandinam | Rural |
| SO29I | 187 | Montgomery | Included entire town but mainly rural |

**The figures relate to the number of mapped species per tetrad (i.e. 2 km × 2 km square) recorded for the Flora of Montgomeryshire (Trueman *et al.*, 1995). The figure excludes the 184 commonest species in the vice county, which were not mapped.**

these sites include the fact that well-tested criteria for designation can be used. Once designated, urban planners are usually quite good at giving protection to such sites, even second and third tier sites. The main problems, as in the countryside, are the usual twin evils of intensification and neglect. Although agricultural intensification is often less of a problem than in the countryside, pressures for development are more likely, and even excessive public interest (nature for *too many* people, most of whom are more interested in cycling, dog walking, jogging etc.) can be very damaging. On the other hand neglect leading to succession – loss of fences, no grazing, desultory cutting, emotive issues with cutting down any trees whatsoever, fly tipping etc. – can be even commoner than in the countryside.

The best examples are either wardened nature reserves or kept private. There is however now often a problem with local knowledge (or ANY remaining knowledge) about non-intensive agricultural and or forestry management. Sometimes also the local authority workforce is dominated by horticulturalists with a very different philosophy of site management, and new staff are often countryside managers who sometimes seem to have been trained mainly in the traditions of game-keeping. There is a real need to school reserve wardens and rangers in the traditional practices of farming and their relationship with nature conservation value. These traditional practices are often at the root of the biodiversity interest.

In addition, urban farms and other schemes less oriented towards nature conservation require significant ecological and nature conservation inputs into their management objectives in order to increase overall urban biodiversity. In particular, Local Authority landowners need to do a proper cost-benefit analysis on letting grazing. Good graziers should get it free, bad ones should pay a very high price. It also still seems to be the case that whilst it is easy to get funds for tree planting and even after-management it is very hard to get any for managing other habitats,

particularly grasslands. Typically also local authority conservation officers have too little influence on site management contract specification.

*Creative conservation* is probably a special case of the above. Some think that imitation hay meadows and ancient woodlands are inappropriate in the urban context, but they are not really any more artificial than flower beds or mown lawns and they do (sometimes) have the advantage of supporting a wider range of species and (occasionally) of looking quite attractive. Whatever the merits the issue is usually again long-term management, although there is potentially a very big issue about 'officious' habitat creation in inappropriate sites. Like every other kind of development habitat creation requires detailed prior site assessment for existing nature conservation value. The hybrid created habitats, with mixed natives and ornamentals, of the Dutch Heemparks and what are possibly their descendants from the Sheffield Landscape School (Dunnett, 1999), must surely present particular problems of management. It is reported that the Heemparks are very expensive to manage, especially in terms of labour.

At the other extreme to the created and semi-natural classical habitats, are those habitats resulting from the natural processes of invasion and succession after site demolition and clearance which can be such a feature of the urban landscape. Gilbert (1989) made the case for these spontaneous successional communities to be allowed to go through the early stages without hindrance. Beyond the obvious need to prevent premature landscaping of such sites and to protect them from abuse, the conservation issue is probably a landscape issue. Long-term survival of the species of these early successional communities probably does not depend on the conservation of the individual site, which would be futile anyway without programmes of repeated disturbance. The crucial factor must be the presence at any one time of sufficient examples of such sites which can regenerate the landscape seed bank for the next generation of examples. Ruderals with their abundant seed production, non-specific seed dispersal and long-lived seed banks would seem to be well adapted for survival under these conditions, but how important is connectivity between sites in the urban landscape? It stands to reason it IS important (Dawson, 1994), but the URGENT investigations (Austin & Angold, 2000) could not demonstrate any positive effect of the proximity of key linear features – canals, railways and rivers – on site plant assemblages. One problem may actually be recognising a corridor when you see one. Private gardens are perhaps part of connectivity and also examples of seed-generating foci. Also allotments and other examples of frequent cultivation need to be included in the assessment.

A landscape approach to urban nature conservation evaluation has started to develop (Young & Jarvis, 2001) and should be encouraged. It offers the possibility of recording spatially all the potentially positive elements in an urban landscape (Plate 1), allowing frequency, proportion, proximity and change to be monitored at the landscape level. The conservation of woodland which arises through these processes of succession also needs to be addressed. Such sites require an unusual length of time to elapse before they mature and they are therefore uncommon. They usually include an array of native and alien species and therefore cannot be easily matched with the National Vegetation Classification. They are often unscheduled and unprotected and require further study.

**Plate 1: A whole landscape approach to evaluating urban nature. A landscape evaluation habitat map for an area of the West Midlands, which can be used to simulate the effects of factors such as redevelopment, habitat creation, change of land use on diversity, connectivity etc.**

**Plate 2: A West Midlands example of an 'ancient' post-industrial site. This nineteenth century blast furnace slag heap and associated wetland in Bilston carries a rich vascular plant, lichen and bryophyte flora including winter annuals such as** *Cerastium semidecandrum* **and** *Aira caryophyllea*, **mire species such as** *Triglochin palustris* **and four species of** *Sphagnum* **moss.**

Garden evaluation is another important developing topic particularly relevant to the maintenance of a rich urban environment. Possibly a nationally or internationally accepted classification of garden vegetation and garden types needs to be developed to aid this assessment.

Somewhere in between the classical sites and the successional sites are what are possibly the most characteristic urban sites. Anyone who has undertaken urban botanical recording would recognise them. Often rich in species, with regionally or even nationally scarce species often present, the vegetation is typically unusual and possibly recombinant (Barker, 2000) without any exact model in the countryside. The conditions are often such that succession is proceeding sufficiently slowly for 'quality' assemblages to develop and therefore also for the feasibility of conserving the individual site to be contemplated. In the West Midlands these are characteristically post-industrial sites, developing particularly on industrial waste (Plate 2) but also in quarries and on railway lines.

What facilitates the development of interest? Repeated or even continuous disturbance may play a part. Stress such as nutrient poverty, drought, phytotoxicity, and mineral imbalance is clearly crucial. There are also interactions: it is very clear that lead-zinc mining spoil inhibits root development. Tests for heavy metal tolerance (Wu & Antonovics, 1976) depends on this effect. The result is that the vegetation on such spoil is chronically susceptible to drought.

Most of these controlling factors are subject to amelioration by succession but the process is often very slow because the conditions are extreme. This impedance of succession is probably at the root of their ecological interest. Possibly toxicity and extreme mineral imbalance are the factors least susceptible to succession so they may be paramount factors (Plate 3), although high (rarely very low) pH may also often be crucial.

**Plate 3:** *Monotropa hypopitys* **(Yellow Bird's-nest). The only known West Midlands county record of this species is on an abandoned and heavily polluted gasworks site.**

**Plate 4:** *Parnassia palustris* **(Grass-of-Parnassus). Parnassia, growing on Solvay process spoil in Upper Silesia in Poland, along with a rich calcicole flora comparable with that found on similar waste in Cheshire and Greater Manchester.**

Some old post-industrial sites have long been celebrated for their floristic richness (Davis, 1976; Greenwood & Gemmell, 1978) and a few are managed as nature reserves, but all conurbations include large and small examples. Existing management is usually serendipitous since such sites are not usually farmed or managed for amenity. Management often includes grazing by rabbits, invertebrates, tethered ponies, Canada Geese. Also fire, trampling and other forms of compaction play a part. Where these factors do not themselves threaten such sites they are often lost to development, especially with the current pressure for cities to develop their brown field sites rather than expanding into the countryside.

The specific issues for conserving these old post-industrial sites include:

- Characterising quality. This is necessary in order to develop an objective system for evaluation against sound standards and hence for site and species protection where appropriate. Ratcliffe's criteria are not easy to apply. The sites are not recognised as being natural (in many ways they are!) and they are not typical, although Rodwell's communities of open habitats (Rodwell, 2000) and those urban Floras with a strong ecological dimension (e.g. Shepherd, 1998) deserve attention as starting points for a new taxonomy. We have still not reached the first stage

in assessing urban sites: we still need a reliable, time-limited set of botanical records for all the UK urban areas. There is still a lot of fundamental recording to be done, and a great deal of analysis of data which is already available (parallel with the study of the city of Plymouth by Kent *et al.*, 1999).

- It is also necessary to undertake more comparisons across the country and between countries (Trueman *et al.*, 2001) (Plate 4) The urban environment has much in common wherever the city and by such studies we will continue to develop a common scientific understanding. It is also important to monitor designated sites in order to understand succession, both to inform programmes of management and to give valid advice on sustainable urban development. The urban Development Corporations, the earlier 'Enterprise Zones' (Plate 5), and now the Regional Development Agencies, organisations designed to power urban change, still sometimes seem to see nature conservation as merely an unimportant obstacle against which they rail petulantly, seemingly seeking to circumvent or ignore what little legislative protection exists. Newspaper headlines about homes/jobs/future threatened by '*a few newts*' are certainly currently on the increase in the West Midlands.

**Plate 5: Merryhill Farm, Dudley. The incorporation of this site close to Dudley town centre into the Dudley Enterprise Zone in the 1980s led to its replacement with a huge shopping centre which competes with the existing town centres and will require massive restructuring of the local transport system.**

- Keeping pollution in perspective and understanding its influence. Can we justify taking a conservation line in the face of environmental pollution? Sometimes such sites have such historical and ecological interest that a very strong case can be made for conservation (Quinn, 1988, 1992). On the other hand the monetary value of the land subsequent to reclamation can be very high in urban areas. Developers seem to think nothing of replacing several metres depth of contaminated land. Despite the development of proposals for the incorporation of nature conservation into urban development (Douglas & Box, 2000) we still seem to need sounder cost-benefit analysis and deeper discussions between planners, developers and ecologists (Harrison & Davies, 2002) when development occurs. These conversations need not be confrontational. Nature conservation may actually be the cheapest and most appropriate end-use for some areas of some sites and might lead to more sustainable and more truly modern developments where wildlife is genuinely incorporated and fostered. We cannot even be sure that the pollution problems would be solved finally by

reclamation: disturbance can mobilise heavy metals, the reclamation may be cosmetic and encourage inappropriate use, and the contaminated product still has to be disposed of somewhere. New EU legislation will require disposal in designated sites and will increase reclamation costs dramatically.

- Giving the application of nature conservation principles to post-industrial sites a human justification. The correlation of nature conservation with conserving industrial heritage has often saved sites. Local groups can be in an ideal situation to identify and celebrate these connections. They need to be encouraged and helped by professionals, but it is not realistic to expect such groups, which often arise in response to a specific threat to a particular site, to be available indefinitely to manage and protect their local sites. Today the old industrial population is dying out and the memory of the industrial past may be getting dim. New human communities are arising with roots elsewhere. The wider relationship between social and ecological value of urban sites, such as the effects on mental and physical health and well-being, needs to be investigated in depth (Niemela, 1999).

- Keeping a valid nature conservation perspective in urban environmental improvement. There is (rightly) a great deal of money available at the moment for improving the urban environment for people and many of the organisations which have sprung up to utilise this resource do not have urban nature conservation as a significant objective. We must not allow urban landscape improvement to lose a wildlife conservation dimension. Ecological appraisal needs to be included in such schemes just as much as in industrial development schemes.

- There must be a degree of scientific objectivity in finding the balance between nature conservation evaluation and social evaluation in the urban landscape. This should apply equally to obsessive conservationists and to persuasive landscape designers who are starting to return to their old pre-Fairbrother (1970) beliefs that they can always improve on nature.

## Conclusions

In summary, experience suggests that urban nature conservation needs to go well beyond the preservation of examples of trapped countryside. If there is a single overriding issue it is the development of relevant ecological expertise and its involvement in urban planning, landscape design, land management and in the celebration of the city.

## Acknowledgements

Thanks are due to John Box, Eleanor Cohn, David Haslam, Ali Glaisher, Peter Millett, Chris Parry and Simon Phipps for useful comments on an early draft.

## References

Austin, K.C. & Angold, P.G. (2000) Influence of landscape components on species recruitment in cities. *Aspects of Applied Biology*, **58**, 115–122.

Barker, G. (ed.) (2000) *Ecological recombination in urban areas: implications for nature conservation*. Proceedings of a workshop held at CEH (Monks Wood). ISBN 1 85716 542 X.

Dawson, D. (1994) *Are habitat corridors conduits for animals and plants in a fragmented landscape?* English Nature Research Report, **94**, Peterborough.

Davis, B.K.N. (1976). Wildlife, urbanisation and industry. *Biological Conservation*, **10**, 249–291.

Douglas, I. & Box, J. (2000) *The changing relationship between cities and biosphere reserves*. A report of the Urban Forum of the UK Man and the Biosphere Committee UK-MAB.

Dunnett, N. (1999) Annual Meadows. *Horticulture Week*. February 1999.

Fairbrother, Nan (1970) *New lives, new landscapes*. Architectural Press, London.

Gilbert, O.L. (1989) *The Ecology of Urban Habitats*. Chapman and Hall, London.

Greenwood, E.F. & Gemmell, R.P. (1978) Derelict industrial land as a habitat for rare plants in S. Lancs and W. Lancs. *Watsonia* **12**(1), 33–40.

Harrison, C. & Davies, G. (2002) Conserving biodiversity that matters: practitioner's perspective on brownfield development and urban nature conservation in London. *Journal of Environmental Management,* **65**, 95–108.

Kent, M., Stevens R.A. & Zhang, L. (1999) Urban plant ecology patterns and processes: a case study of the flora of the city of Plymouth, Devon, UK. *Journal of Biogeography*, **26**, 1281–1298.

Niemela, J. (1999) Ecology in urban planning. *Terra*, **111**(3), 157–64.

Montgomeryshire Field Society and Montgomeryshire Wildlife Trust, Welshpool, UK.

Quinn, M-L. (1988) Tennesee's Copper Basin: a case for preserving an abused landscape. *Journal of Soil and Water Conservation* **43**(2): 140–144

Quinn, M-L (1992). Should all degraded land be restored? A look at the Appalachian copper basin. *Land degradation and rehabilitation* **3**: 115–134

Rodwell, J.S. (ed.) (2000) *British Plant Communities Volume 5: Maritime Communities and Vegetation of Open Habitats*. Cambridge University Press, Cambridge.

Shepherd, P. (1998) *The Plants of Nottingham*. Wildtrack Publishing, Sheffield.

Trueman, I.C., Morton, A & Wainwright, M. (1995) *The Flora of Montgomeryshire*. The Montgomeryshire Field Society/The Montgomeryshire Wildlife Trust, Welshpool.

Trueman, I.C., Cohn, E.V.J.C., Tokarska-Guzik, B., Rostański, A. & Woźniak, G. (2001) Calcareous waste slurry as wildlife habitat in England and Poland. In Sarsby, R.W. & Meggyes, T. (eds) *Proceedings of GREEN 3: the Third International Symposium on Geotechnics related to the European Environment, Berlin 2000* Thomas Telford UK.

Wu, L. & Antonovics, J. (1976). Experimental ecological genetics in *Plantago II*: lead tolerance in *Plantago lanceolata* and Cynodon dactylon from a roadside. *Ecology*, **57**, 205–208.

Young, C.H., & Jarvis, P.J. (2001. Measuring urban habitat fragmentation: an example from the Black Country, UK. *Landscape Ecology*, **16**, 643–658.

*International Urban Ecology Review*, 5, Urban Environments Issue

# *Studying wildlife distribution using 'citizen' science public sightings confirms how suburban deer are now found throughout the United Kingdom*

## Ian D. Rotherham[1] and Mark Walker[2]

[1]Professor of Environmental Geography, Sheffield Hallam University; [2]Work experience placement, HEC Associates, Sheffield

## Abstract

Engaging public participation in wildlife research, popularly termed 'citizen science', is currently much in vogue. However, it is often the case that scientifically useful information is not generated through such projects. In order to test this idea, we used publically submitted data to study the nationwide extent of suburban deer. The long-term Sheffield and Peak District regional studies have used a citizen science approach since the 1980s (Rotherham & Derbyshire, 2012), well before the current vogue for the term or the availability of electronic networking and social media. Locations from where sightings were submitted were studied with aerial photographs and categorized according to habitat, land use and dwelling number. Deer were found to live in suburban areas nationwide, with roe being the most commonly seen. This innovative use of aerial images alongside the submitted records suggests a method to enhance the practical use and application of 'citizen science' records. Although for many decades, deer have been known to live in urban and peri-urban settings, (see for example, Rotherham *et al.*, 1996), further local and national study is needed.

Key words: urban deer, BTO Garden Bird Surveys, UK national distribution

## Introduction

According to Ward (2005), it is often stated that wild deer in Great Britain are increasing in range and number. The 2005 paper presented the first quantified estimate of range expansion for all six UK species at a national level. Working with British Deer Society

**Red deer © Peter Wolstenholme**

database it was possible to compare national surveys of deer presence in 10 km squares between 1972 and 2002. This study generated the following results: 1) red deer range was estimated to have expanded at a compound rate of 0.3% per year, 2) fallow deer was at 1.8%, 3) Chinese water deer was at 2.0%, 4) roe deer was at 2.3%, 5) Japanese sika was at 5.3% and 6) Reeves' muntjac was at 8.2%. Furthermore all species were expected to expand their range over the foreseeable future. Roe deer was the most widespread and it was predicted that within ten years it would be present in 79% of all 10 km squares in mainland Britain. Ward noted that deer range expansion brought both benefits and costs. Additionally, it was concluded that management would be necessary to control both environmental damage and deer suffering. The study by Ward (2005) did not address urban deer specifically, but the findings provide context and background for work on deer in urban and peri-urban environments. Additionally, the potential problems, (as noted by McCarthy *et al.* (1996), increase significantly once deer advance into more urbanised landscapes.

However, although isolated anecdotal reports and regional studies exist, there has been no systematic examination or quantification of the extent to which deer inhabit suburban areas at a UK national level. Indeed the extent to which deer are established in suburban areas in the UK is not fully known. In order to glean more information on this phenomenon, aerial photographs were studied along with data on deer sighting reported as part of the BTO Garden BirdWatch survey. Areas where deer were seen were categorised according to habitat, land use and dwelling number. Unsurprisingly given previous observations and regional studies such as around Sheffield, suburban deer were discovered to be a common occurrence. Roe deer *Capreolus capreolus* were most frequently observed. A large proportion of sightings were in heavily residential areas. The proportion of urban deer reports from the BTO survey remained steady over ten years. This innovative use of aerial images suggests an additional method for

others to apply in order to examine geographical records and wildlife datasets. Furthermore, as suggested would happen in the 1990s (McCarthy *et al.*, 1996), whilst traditionally considered as rural countryside animals, deer should now be firmly considered suburban or peri-urban dwellers. It has already been noted that measures to mitigate possible negative impacts of urban deer should be considered where appropriate. These should include the incorporation of road passes where needed to avoid road traffic accidents (RTAs), and in some cases control of deer populations. More thoughtful urban and peri-urban landscaping schemes should also take into account the urbanising deer phenomenon. In urban and peri-urban areas, as forecast by McCarthy *et al.* and Rotherham (1996), the latter remain contentious and problematic.

Encouraging public participation in wildlife research is currently much in vogue, and has been popularised with the name 'citizen science'. Silverton (2009) defines 'citizen scientists' as volunteers engaged in active inquiry through data collection and processing. Such volunteers offer an invaluable cost effective and often knowledgeable resource (Cohn, 2008). Many, particularly urban, people have become disconnected with the natural environments, a phenomenon known as 'cultural disconnection' (Rotherham, 2009). 'Citizen Science' projects offer the possibility to reconnect communities with wildlife. However, it is often the case that scientifically useful information fails to be generated through such projects (Cohn, 2008). How can such publicly submitted observations be used to study wildlife conservation and management problems?

Within the United Kingdom reports of urban living deer have been reported over many decades (McCarthy & Rotherham, 1996; Jones *et al.*, 1999; Rotherham, 2001). Reports of red deer entering suburban areas of Sheffield were made from the 1980s onwards (McCarthy & Rotherham, 1996). In Glasgow, roe deer live within the boundaries of the urban areas and have been the subject of considerable study in part inspired by the earlier Sheffield research (Dandy *et al.*, 2009; Dandy *et al.*, 2011). Perhaps the presence of deer in such areas should come as no surprise. Urban and suburban dwellings now comprise 7% of land use in Britain, and 11% in England (Watson & Albon, 2010). For urbanising deer, suburban areas are predator free, lack hunting threats and offer plentiful food supplies, and often increasing tree cover with linear planted greenways. Similarly in the United States white-tailed deer (*Odocoileus virginianus* Zimmermann 1780) have long been an issue in urban situations (e.g. Nielsen *et al.*, 1997; Rondeau & Conrad, 2003).

Great changes have taken place in the composition of the British landscape during the twentieth century with the human population increasing substantially. Suburbanisation, the spread of human habitation over great areas of land adjoining cities and towns, has occurred. Urban and suburban dwellings now comprise 6.8% of land use in Britain, and 10.6% in England (Watson & Albon, 2010). A 'suburb' is defined as 'the outlying district of a city, especially as residential one' (OED, 2015). This definition places emphasis on the residential nature of suburban areas, with human housing dominating this land category. Deer are opportunist animals and readily colonise suitable habitat. Suburbia offers a potentially rich resource for deer; with plentiful grazing, sufficient cover, and a lack of hunting pressure.

Following earlier decimation populations of British deer have increased substantially in recent years (Ward, 2005). This increased abundance, allied to their opportunist nature means deer would be thought to readily colonise this newly available habitat.

Although there is some anecdotal evidence that deer do enter suburbanised settings (McCarthy *et al.*, 1996; Dandy *et al.*, 2011) the extent of this phenomenon is unknown and has never been quantified beyond local

and regional case studies (e.g. Rotherham, 1996, 2000, 2001, 2003, 2008, 2009; Rotherham *et al.*, 1996; McCarthy & Rotherham, 1993, 1996). Whether deer now live in UK suburban areas dominated by residential housing has never been more widely assessed or quantified. Whether previous reports are simply isolated occurrences in particular towns or an indication of national trends is unknown. This is pertinent as increases in deer abundance or distribution is likely to lead to increased deer-human interactions. Deer mediated road-traffic accidents; through collision or driver swerving, can cause substantial damage (Deer Initiative, 2007; J. Langbein pers. comm.). In 2010, the Office of National Statistics recorded 150 accidents, eleven of which were fatal, as being caused by animals in the roadway (ONS, 2010). Deer also help spread diseases such as Lyme disease *Borrelia burgdorferi* (Johnson *et al.*, 1984; Simpson, 2002) and damage gardens and crops (Putman & Moore, 1998). Therefore ascertaining the extent that deer dwell in suburban areas is pertinent.

Suburban deer present a number of potential management issues. The main problems of deer in urban and peri-urban areas are deer-vehicle collisions (Porter *et al.*, 2004; RDS, 2006), but also include damage to gardens and to wildlife habitats such as woods. In 2010, the Office of National Statistics recorded 150 accidents, 11 of which were fatal, as being caused by animals in the roadway (ONS, 2010). Deer may also contribute to the spread of diseases such as Lyme disease *Borrelia burgdorferi* (Johnson *et al.*, 1984; Randolf, 2001). Deer also cause damage to gardens, horticulture, and farm crops (Putman & Moore, 1998). However, deer are also charismatic and impressive animals and are generally welcomed by local residents – at least until damage becomes apparent. They offer the potential to engage local communities with their local environment (Rotherham, 2009). It seems that in urban and peri-urban environments deer are generally perceived positively by local people (Ballantyne, 2012).

In recent years, aerial photographs of specific locations have become increasingly available. Such images allow detailed features of an area, such as individual houses, fields or even cars, to be ascertained. However, the use of such photographs has yet to be used extensively for wildlife research purposes. Since 1995, the British Trust for Ornithology has organised an annual 'Garden BirdWatch', with participants submitting garden bird sightings. Other wildlife observed is recorded in addition to avian sightings. These reports provide grid references for each report submitted. These records provide a source of data for examining a range of wildlife questions using aerial photography. We used this dataset to ascertain the habitat make-up of areas where deer sightings were made and thus assess the extent that deer occur in suburban or urban areas. This was an innovative use of newly available aerial photographs allied with the excellent BTO dataset to assess a previously unquantified phenomenon. This study shows how existing datasets can be used in different ways to generate new knowledge.

Although a common phenomenon, the nationwide extent of suburban deer has not been specifically studied in the UK. In this context, the use of publically submitted 'Citizen Science' records offers the potential to study the distribution of suburban deer across the United Kingdom. We examined one database, the BTO Garden BirdWatch dataset, to consider the UK nationwide extent of suburban deer. On-line aerial photographs were then used to determine the habitat make-up from where sightings were submitted. This innovative approach, suggests an approach to answer wildlife questions using openly accessible records.

## Methods

Records of deer submitted to the BTO Garden BirdWatch scheme were examined. Although surveying garden birds, respondents are asked to record other garden wildlife present during such surveys. Records containing deer sightings spanned twelve years; from the 1st of December 2002 to 30th November 2014.

Interestingly, the surveys presented 10,970 records with sightings of deer, and these were from 615 separate locations. The records gave information on the species and number of deer seen, plus an indication of the geographical location through a six-figure grid reference. It has to be noted that since these were records submitted by an interested public rather than from specialists and so the precise identification may not be one hundred percent reliable. However, considerable experience with the Sheffield and Peak District surveys over around thirty years indicates that most public records are good. Any errors usually creep in with confusion between red deer hinds, fawns and followers which are often mistaken for roe. This does need to be considered in any final analysis of the citizen science data.

*Habitat Analysis*: Using the six-figure grid reference provided an aerial photograph was obtained for the approximate location of the site. Images were obtained from www.finda-gridreference.com which allows examination of overhead views of specific locations. Six-figure grid references are accurate to a distance of 100 m. An aerial image of 1 $km^2$ centred on these grid references was examined to provide a realistic impression of the composition of the surrounding environment. This was used to decide on the approximate habitat composition of the area.

*Levels of Suburbanisation*: Firstly, aerial images were assessed and the presence of 'suburban' areas of housing was determined in each km2. A suburban area was defined as being an agglomeration of several houses centred along a roadway, often being adjacent to a town. Secondly, the number of residential dwellings was estimated from each aerial image on a scale from 0 to 30 plus. If squares contained only farm buildings, or no obvious dwellings at all, this was also noted. As suburbia, by definition, is comprised of residential housing, ascertaining the number of houses is relevant.

*Habitat Type*: Firstly the main habitat type was determined at each location. Land use was classified into woodland, pasture or grassland, arable farmland, suburban housing, industry, water such as sea or lake, heathland or moorland, and mountain.

Secondly, the proportion of each area containing each habitat type was estimated. This was ranked for analysis purposes. The habitat type covering the largest proportion of an area was assigned a rank of 1, the second largest habitat was assigned a rank of 2, *etc*. Where habitats appeared to cover equally large areas, joint ranks were given. The presence of a motorway or major trunk road was in a square was noted.

Mapping of sightings: Deer sightings were mapped using DMAP distribution mapping software for Windows 7.0 onto a British map with 10 km$^2$ squares. Maps were produced for all deer sightings, those reported in areas classed as being 'residential', and in those having more than 30 or more residential houses. The percentage of sightings made in areas with residential areas was calculated for the years 2002 to 2014.

# Results

Results show that deer were reported from many suburban gardens. There were 10,979 reports containing deer were obtained from the BTO covering the time period from 2002 to 2014. Of these, 615 were submissions from single locations, and overall, the data indicate a nationwide extent of deer in suburban situations. Table 1 (below) presents the number of each species reported. The most commonly seen deer species in gardens was

roe deer *Capreolus capreolus* L. 1758, the least frequently seen was sika deer *Cervus nippon* Temminck 1838. Fallow deer *Dama dama* L. 1758, were the second most frequently seen species, followed by red deer *Cervus elaphus* L. 1758. Chinese Water Deer *Hydropotes inermis* Swinhoe 1870 and Reeve's Muntjac *Muntiacus reevesi* Ogilby 1839 were only seen in small numbers.

Table 1 shows the number of deer reported in areas comprising a 'suburban residential' estate. There were 51% of sightings in suburban or residential areas and 49% from non-suburban areas.

Figure 1 (page 72) shows the number and percentage of sightings from areas with different numbers of houses. This indicated 40% of sightings occurred in areas with no housing or only farmhouses. However, 56% of sightings were made in areas with housing, 25% of areas where deer were sighted contained over 30 houses and thus were considerably suburbanised.

Deer were seen in a many suburban, residential or semi-urban locations. Housing was assessed as being the dominant land use in 15% of the 601 records where land use could be examined (Figure 2, page 72). The most dominant use of land where deer sightings were made was pasture or grassland which was dominant in 41% of areas. Arable

**Table 1: The number of each deer species reported and the number of 1 km$^2$ areas around reported sightings which possessed a 'suburban residential' estate or area**

| Table 1: Reported sightings from the survey | | |
|---|---|---|
| **Species** | **Number of Reported Sightings (% total)** | **Number of areas with 'residential' estate (% of species sightings)** |
| Roe Deer | 506 (82.8%) | 254 (50.19%) |
| Red Deer | 38 (6.21%) | 13 (34.21%) |
| Fallow Deer | 52 (8.5%) | 24 (46.15%) |
| Chinese Water Deer | 9 (1.47%) | 4 (66.67%) |
| Sika Deer | 6 (0.98%) | 3 (33.3%) |
| Muntjac | 0 (0) | 0 (0%) |

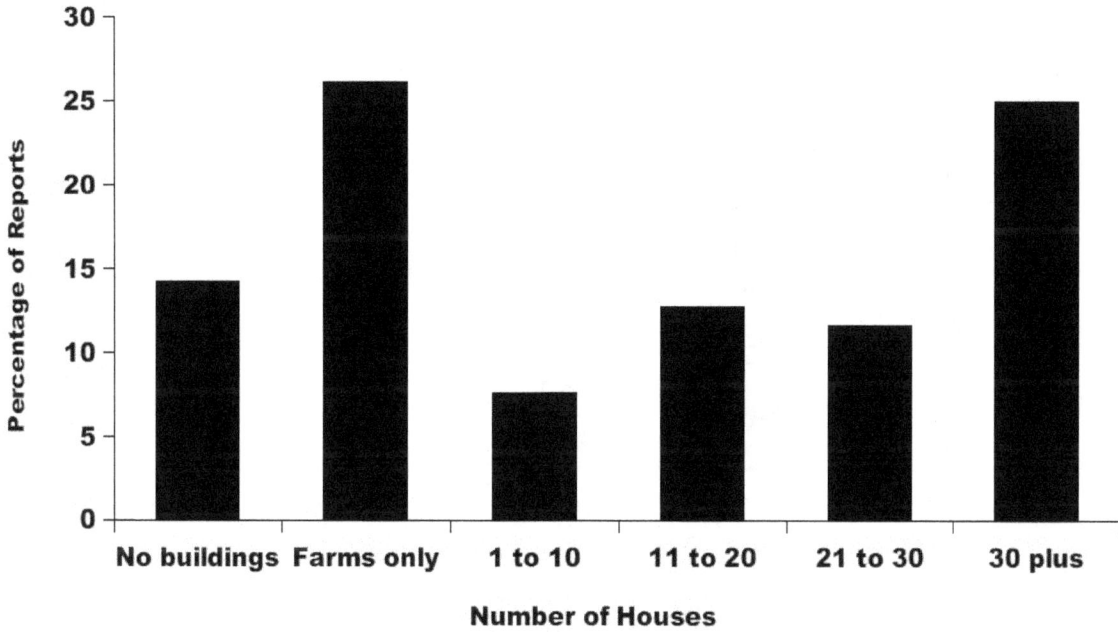

Figure 1: The number of houses ascertained from aerial photographs in each 1 km$^2$ area where deer were reported in the BTO GardenWatch survey, figures as percentages

Figure 2: The main habitat type in each 1 km$^2$ square area where deer reported in gardens

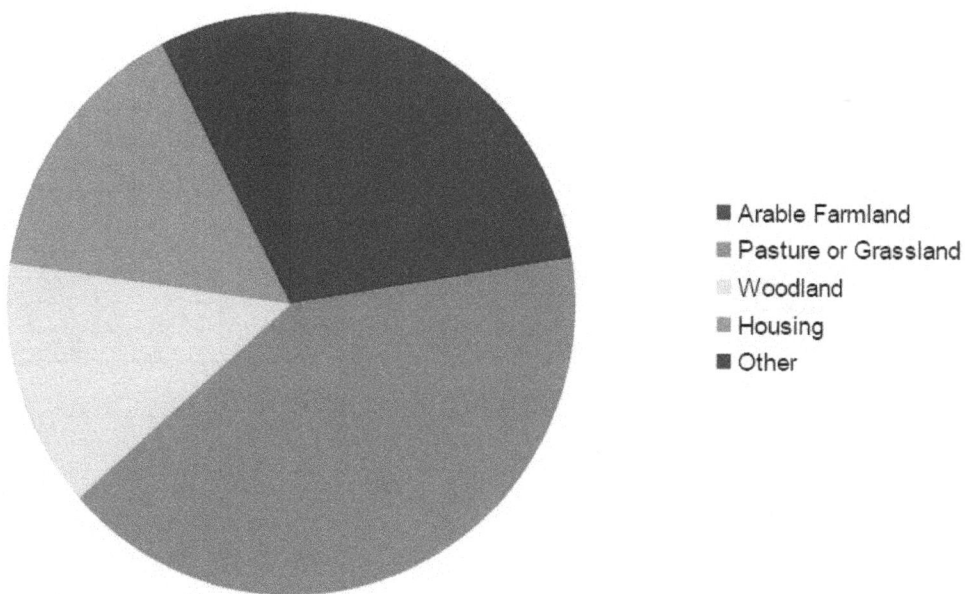

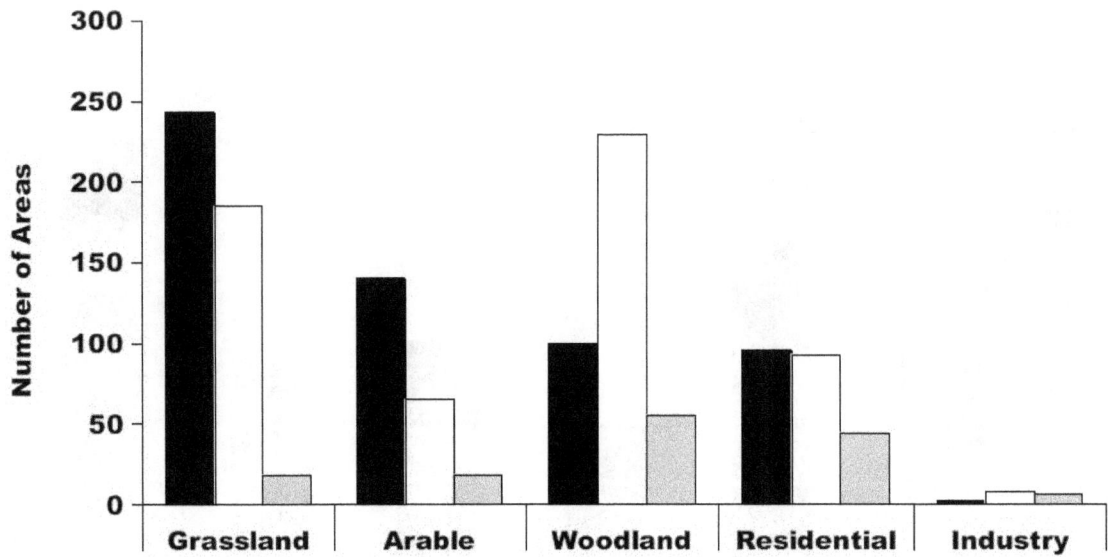

**Figure 3: The ranking of different habitat type in each examined 1 km$^2$ area around reported sightings. Key: Black bars ranked as one, clear bars as 2 and grey bars as 3**

**Figure 4: Distribution of deer sightings across the UK: a) location of deer sightings within 10 km squares; b) deer sightings in areas classed as possessing a 'suburban residential' estate; c) deer sightings from areas with 30 or more residential dwelling houses**

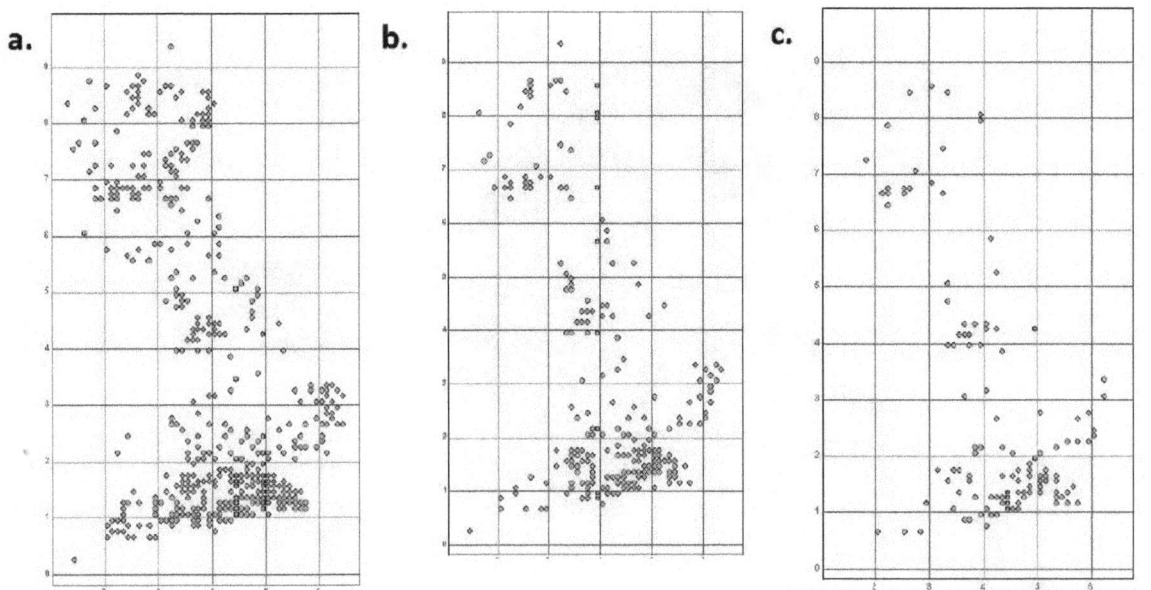

land use was dominant in 23% and woodland in 14%. Figure 3 shows the number of times each habitat was ranked as 1, 2 or 3 in each area (Figure 3, page 73).

From the dataset, 43 (7%) of sightings were from areas which had motorways or major roads. 26, (4% total) of these were in residential areas. Of those with a motorway, 38% were in suburban residential areas, and 62% in non-suburban areas.

*Distribution and trends*: Mapping of all deer reported sightings showed that sightings were well distributed across the British Isles (Figure 4a, page 74). The distribution of sightings reflected human population distribution, with a concentration in southern England. Sightings of deer in suburban areas (Figure 4b, page 74), those areas possessing a suburban residential area or in excess of 30 houses (Figure 4c, page 74), were likewise well scattered across Britain but exhibited a similar southern concentration. The percentage of sightings in areas with suburban residential estates was relatively constant between 2002 and 2014.

# Discussion

Deer are clearly well established in suburban areas, a situation predicted by McCarthy *et al.* (1996). The results provide some of the first quantitative evidence specifically of the widespread use of suburban areas by deer. Roe deer were most commonly reported, probably due to its smallness and innocuous lifestyle. Larger species such as red and sika deer were seen much less often. However, that some sightings were made close to human habitation is of significance, given their ability to range over large areas and possibly enter suburban areas that this ability provides. Surprising is the relative paucity of sightings of muntjac, which given its size would be thought to be common in gardens. The proportions of sightings for each species in suburban areas is interesting as they show more roe deer were reported in suburban or built up than rural areas, but this is probably a reflection of under-recording in the rural zone, its ease of penetration into peri-urban and urban areas, and the ease with which it is seen these situations. Muntjac on the other hand is relatively small, rather nocturnal and secretive in habits.

**Roe deer ©Peter Wolstenholme**

Many sightings came from areas with suburban residential estates and housing was the largest land use for a substantial number of sightings. This also remained a major land use in a large number of areas where other uses dominated, often being ranked as the second or third major land use; clearly related to this being a garden-based census. A high proportion of deer were sighted in areas with conversely only farms or no housing, and areas with 30 plus houses. By its very nature, data contained within the BTO Garden BirdWatch scheme is likely to reflect wildlife presence in semi-urban or suburban settings. The scheme is promoted as a garden BirdWatch, thus participants are likely to be biased to living in suburban areas. That most sightings occurred in areas possessing suburban residential estates and where housing made up more than 50% of the land use indicate reports were predominately from suburban districts.

The presence of motorways in many squares with sightings is particularly interesting as they may provide conduits for deer; funnelling them into urban areas and allowing them to penetrate along corridors to the heart of urban areas. This is particularly so when new arterial road infra-structures and was predicted by McCarthy *et al.* (1996).

Many previous studies have concentrated on urban deer in particular local areas (e.g. Dandy *et al.*, 2011; McCarthy & Rotherham, 1996). This work shows that suburban deer are not limited to one area but this urban penetration is a truly nationwide phenomenon. Although strongly concentrated in southern Britain, this likely reflects either the human or deer population distribution which are both heavily concentrated in the south. It may indicate that deer are more urban in the south where urban sprawl is of greater extent. There will also be a bias dependent of the geographic spread of the bird survey recorders.

Importantly this investigation illustrates a possible method to use publically submitted wildlife observations to answer scientifically relevant questions and guide wildlife management. Although it is easy to disparage observations made by non-professional naturalists, the public offer a useful resource for wildlife research (Cohn, 2008). Studying national deer distribution is just possible application of using such public records. Often, a single individual or small group of scientists could not generate data needed for truly nationwide study. Some previous surveys of deer, such as the British Deer Society Deer Survey, utilised only society members for such work and generated much smaller datasets (REF), but of course these are by self-selecting and knowledgeable individuals.

*Limitations*: There are a number of limitations. Members of the public may be unreliable sources of wildlife observations due to misi-dentification. However, since deer generically are distinctive members of the British fauna misidentification is unlikely for this broader group though species confusion may occur. The smaller species such as muntjac are likely to be under-recorded in surveys reliant on sightings rather than calls. The weight of ob-servations used is likely to ensure that general trends are interpreted correctly.

Categorisation of habitats was often subjective. Suburban housing for example contains gardens with grass and trees, which are features of other habitats. In so-called urban areas, 54% of land is actually 'greenspace' comprised of allotments, verges and parks (Watson & Albon, 2010). Identification of land types was probably not always correct; for example arable fields could easily be mistaken for pasture. The use of publicly generated observations poses problems; members of the public may misidentify animals, or false observations may be submitted.

The results probably underestimate urban deer. Direct counts are well known to significantly underestimate deer numbers (Anderson *et al.,* 2013). Deer were not the primary aim of surveying; and thus many respondents may not have listed deer sightings. Although it might be suggested that being relatively large deer could not be overlooked, their secretiveness allied to nocturnal habits, means they are likely missed in many sites.

An increase in deer abundance in urban and residential areas will lead to increased human-wildlife interactions. Increased risks from road-traffic accidents, increases in garden damage and increased disease spread may occur. Public perceptions of deer are coloured by personal experiences (Dandy *et al.*, 2011). The public are key stake-holders in deer management, thus their attitudes and views are important to determining deer management issues (Messmer *et al.*, 1997). An assessment of habitat type is interesting because this has been shown to be important in determining the number of road-traffic accidents (White *et al.*, 2004).

## Overall conclusions

Despite the limitations, performing the study was extremely useful. This was a first attempt to numerically quantify the extent that deer are dwelling in suburban and peri-urban areas beyond the local or regional. It provides a strong indication that deer in the UK are found in urban and peri-urban areas and that they are well established there. With habitat classification, since the outline of housing estates is unmistakeable, identification of urban areas and residential housing estates was relatively easy through aerial photography. Although the exact location of sightings could not be pinpointed, the distances covered per square (10 km$^2$), are soon crossed by quick-moving animals. Thus deer can easily move from adjoining habitats into urban areas quite speedily and even as a transient use of resources.

Overall, the results probably underestimate urban deer. Direct counts are well known to significantly underestimate deer numbers (Anderson *et al.*, 2013). Deer were not the primary aim of surveying and many respondents may not have listed deer sightings. Although one may believe that deer being relatively large could not be overlooked, their tendency to secretiveness allied to

nocturnal habits, means they are likely missed in many sites. In the case of the red deer now colonising the western suburbs of Sheffield however, this is not the case. The animals are large, obvious, and during the rut, noisy.

Increase in deer abundance in urban and residential areas is leading to increased human-wildlife interactions. Increased risks from road-traffic accidents, in garden damage and associated tick-borne disease may follow. Public perceptions of deer are coloured by personal experiences (Dandy *et al.*, 2011). The public are key stakeholders in deer management, thus their attitudes and views are important to determining deer management issues (Messmer *et al.*, 1997). In this context, an assessment of habitat type is interesting because this has been shown to be important in determining the number of road-traffic accidents (White *et al.*, 2004).

The study shows how new technological advances can be used for wildlife research purposes. Aerial photographs offer the chance to examine reported locations of wildlife at great resolution. This offers the chance to utilise existing datasets for new purposes as done here. The BTO GardenWatch reports are an excellent source of data enabling study of a variety of wildlife topics. In summary, the results suggest that deer are now an established feature of suburbia. Measures to alleviate the possible impact of deer in residential areas should be undertaken. This could include appropriate road signage, public education measures, use of fencing to prevent road-collisions, and carefully planned design of new urban and peri-urban landscaping schemes such as along highways. Many of these ideas were first proposed at the 1996 *Deer or the New Woodlands* meeting in Sheffield (McCarthy *et al.*, 1996), but it remains for them to be widely implemented.

The study demonstrates the potential use of 'Citizen Science' produced records to study wildlife distribution. Additionally, the work confirms the increasingly urbanised nature of deer across the United Kingdom. It indicates how publicly generated data can be used to generate scientific information and used to study population dynamics. The use of freely available aerial imagery offers a cost-effective method to study wildlife on a national level.

## Acknowledgements

Thanks to Dr Peter Lack, Information Services Manager, British Trust for Ornithology, The Nunnery, Thetford, Norfolk for access to the BTO Garden BirdWatch survey data. Anonymous referees of an early draft are also acknowledged

## References

Anderson, C.W., Nielsen, C.K., Hester, C.M., Hubbard, R.D., Stroud, J.K., & Schauber, E.M. (2013) Comparison of indirect and direct methods of distance sampling for estimating density of white-tailed deer. *Wildlife Society Bulletin*, **37** (1), 146-154.

Ballantyne, S. (2012) Urban Biodiversity: Successes and Challenges: Human perceptions towards peri-urban deer in Central Scotland. *The Glasgow Naturalist*, **25**, Part 4

Dandy, N., Ballantyne, S., Moseley, D., Gill, R., & Quine, C. (2009) *The management of roe deer in peri-urban Scotland*. Forestry Research, Edinburgh.

Dandy, N., Ballantyne, S., Moseley, D., Gill, R., Peace, A., & Quine, C. (2011) Preferences for wildlife management methods among the peri-urban public in Scotland. *European Journal of Wildlife Research*, **57** (6), 1213-1221.

Deer Initiative (2007) *Deer on our Roads – Counting the Costs*. Deer Initiative, Wrexham.

Fuller, R.J., & Gill, R.M.A. (2001) Ecological impacts of increasing numbers of deer in British woodland. *Forestry*, **74** (3), 193-199.

McCarthy, A. J. & Rotherham, I. D. (1993) Muntjac (*Muntiacus reevesi* Ogilby) on the Sheffield urban fringe - introduction on natural colonisation? *Sorby Record*, **3**, 3-6.

McCarthy, A.J., & Rotherham, I.D. (1994) Deer in the Sheffield region including the Eastern Peak District. *Naturalist*, **119**, 103-110.

McCarthy, A.J., & Rotherham, I.D. (1996) Urban Deer, Community Forests and control - Roe Deer in the urban fringe - a Sheffield case study. *The Deer*, **10**, 26-27.

McCarthy, A.J., Baker, A., & Rotherham, I.D. (1996) Urban-fringe Deer Management Issues - a South Yorkshire Case Study. *British Wildlife*, **8**, 12-19.

Messmer, T.E.A., Cornicelli, L., Decker, D.J., & Hewitt, D.G. (1997) Stakeholder acceptance of urban deer management techniques. *Wildlife Society Bulletin*, **25**, 360-366.

Nielsen, C.K., Anderson, R.G., & Grund, M.D. (2003) Landscape Influences on Deer-Vehicle Accident Areas in an Urban Environment. *The Journal of Wildlife Management*, **67**, 46-51

Nielsen, C.K., Porter, W.F., & Underwood, H.B. (1997) An adaptive management approach to controlling suburban deer. *Wildlife Society Bulletin*, **25**, 470-477.

Porter, W.F., Underwood, H.B., & Woodard, J.L. (2004) Movement behavior, dispersal, and the potential for localized management of deer in a suburban environment. *Journal of Wildlife Management*, **68** (2), 247-256.

Putman, R.J., & Moore, N.P. (1998) Impact of deer in lowland Britain on agriculture, forestry and conservation habitats. *Mammal Review*, **28** (4), 141-164.

Randolph, S.E. (2001) The Shifting Landscape of Tick-borne Zoonoses: Tick-Borne Encephalitis and Lyme Borreliosis in Europe. *Philosophical Transactions of the Royal Society of London. Series B: Biological Sciences,* **356** (1411), 1045-1056.

Rondeau, D., & Conrad, J.M. (2003) Managing Urban Deer. *American Journal of Agricultural Economics,* **85**, 266-281.

Rotherham, I.D. (1996) Deer or the New Woodlands? - A Local Authority Perspective. In: Deer or the New Woodlands? *The Journal of Practical Ecology and Conservation, Special Publication,* No.1, November 1996. Wildtrack Publishing, Sheffield, 59-63.

Rotherham, I.D. (2000) Deer on the Peak District's urban fringe – a South Yorkshire case study. *Peak District Journal of Natural History and Archaeology,* 2, 75-88.

Rotherham, I.D. (2001) Urban Deer: A South Yorkshire case study. *The Deer,* 11 (10), 566-569.

Rotherham, I.D. (2003) *Deer in urban and urban fringe areas – trends, issues and challenges.* In: Goldberg, E. (ed.) *Proceedings of the Future for Deer Conference, 28th - 29th March 2003.* English Nature Research Reports No. 548, English Nature, Peterborough, 40-48.

Rotherham, I.D. (2008) *Urban Deer: issues and conflicts and the results of a twenty-year study of re-colonisation.* Abstract paper in: Proceedings of the Mammal Society Autumn Symposium on Mammals in Urban Environments, 14-15 November 2008, London Zoo. Unpaged.

Rotherham, I.D. (2009) Deer in the Urban Fringe. *World of Trees,* 18, 46-49.

Rotherham, I.D., McCarthy, M., & Baker, A.J. (1996) *Deer Colonisation of Urban Fringe Areas, with Particular Reference to the South Yorkshire Forest Area, Sheffield.* South Yorkshire Forest, The Forestry Authority, & SCC, Sheffield.

RDS [Rural Development Service] (2006) *Deer: problems in urban and suburban areas.* Technical Advice Note 37, HMSO, London.

Simpson, V.R. (2002) Wild animals as reservoirs of infectious diseases in the UK. *Veterinary Journal,* 163 (2), 128-146.

Silvertown, J. (2009) A new dawn for citizen science. *Trends in Ecology & Evolution,* 24 (9), 467-471.

The Oxford English Dictionary (2015) *OED Online.* Oxford University Press. http://www.oxforddictionaries.com/

Ward, A.I. (2005) Expanding ranges of wild and feral deer in Great Britain. *Mammal Review,* 35, (2), 165–173

Ward, A.I. (2007) Trends in deer distribution and abundance within the UK. Proceedings of the Deer Initiative Conference '*Deer, Habitats and Impacts*', Buxton, UK, March 23rd, 2007.

Watson, R., & Albon, S. (2010) *UK National Ecosystem Assessment: Draft synthesis of current status and recent trends.* UK National Ecosystem Assessment (no location given).

White, P.C., Smart, J.C., Böhm, M., Langbein, J., & Ward, A.I. (2004) *Economic impacts of wild deer in the east of England.* Central Science Laboratory, Woodchester Park, Gloucestershire.

*International Urban Ecology Review*, 5, Urban Environments Issue

# Some early issues in urban nature conservation

**David Goode**
University College London

This paper was originally given at a conference on Urban Environments, History, Biodiversity and Culture at Sheffield Hallam University in April 2005.

## Introduction

I was asked to review the early days of urban nature conservation, to see what we have achieved and then to look ahead to see where we are going. It was a tall order and difficult to know where best to start. But I chose two books which between them go a long way to set the scene. They are *London's Natural History* by Richard Fitter, published in 1945, and John Kieran's *Natural History of New York City*, published in 1959. These two authors had a real understanding of the natural history of cities, and brought their infectious enthusiasm to the subject. They were naturalists who knew the ecology of their cities inside out. Richard Fitter's book was one of the earliest New Naturalist volumes and, such was its appeal, it is still in print today. He provided remarkable insights into the way nature adapts and even thrives within the urban environment. It is a great read. You will find more difficulty coming across the book on New York, but it too is well worth reading. In one passage he has a lovely description of a day in Manhattan among the skyscrapers when he suddenly heard geese honking high in the sky. They were Snow Geese on migration from the arctic to somewhere down on the Gulf of Mexico. His excitement is palpable. The book is full of such incidents. He brought nature to the attention of New Yorkers through his enthusiasm and detailed understanding of ecology. Between them Fitter and Kieran set the scene beautifully for an appreciation of the natural world of towns and cities.

But it wasn't until the 1970s that this appreciation started to find its way into ideas and policies for urban development. Crucial

people involved were Ian McHarg with his pioneering book *Design with Nature* in 1969, and Nan Fairbrother's *'New Lives, New Landscapes'* in 1970. Later came Richard Mabey with 'The Unofficial Countryside', published in 1973, in which he first pointed out the value of abandoned land as urban wilderness. He argued that railway sidings and odd little wastelands really matter with their rich array of habitats that we take for granted, like the swathes of rosebay willow-herb growing on rail-sides, or quite remarkable situations like old railway sidings turning into wonderful pieces of birch woodland full of fungi. It is here that you find real biodiversity in the midst of the urban scene and Richard Mabey was one of the first to point this out.

## Urban Ecology and the Professions

In 1974, a conference was organised in Manchester on urban wildlife by landscape architects, notably Ian Laurie, with the Institute of Landscape Architecture and the Landscape Research Unit (Laurie, 1975). Stimulated by a ground breaking meeting in Washington DC in 1968 the Manchester gathering was instrumental in taking new ideas forward in the UK. As well as landscape architects the conference involved both ecologists and naturalists, together with farmers and land managers dealing with the urban fringe. But it was really led by landscape designers concerned about the paucity of nature in our emerging post-war townscapes. For the first time too there was consideration of human needs in relation to the natural environment of towns and cities.

Another key contribution at that time came from Brian Davis who provided a comprehensive review of wildlife in urban and industrial areas in a paper in Biological Conservation in 1976. His was a lone voice among research ecologists, in recognising the value of man-made habitats for wildlife conservation. At that time I was working for the Nature Conservancy Council in Scotland and I well remember officers of the NCC rubbishing his work on the grounds that real nature was found in the wild landscapes of the Highlands and Islands and we shouldn't waste time on this urban stuff. That was before we heard about the botany of shale-bings. Suddenly NCC officers were getting agitated about shale-bings. These were the slag heaps of the old oil-shale industry which were a feature of the landscape in the central belt between Edinburgh and Glasgow. Being very acidic they were found to have all sorts of strange and unexpected plants growing on them. Some botanists advocated that the bings needed to be conserved. This was my first direct experience of urban nature conservation, and the passions that it aroused, for there were strong antagonisms from the more traditional conservationists.

## The Endless Village

Elsewhere things were happening very quickly. The West Midlands Metropolitan County had recently come into being and was required to produce a structure plan. So they contacted the NCC and asked which sites were important for nature conservation. This request landed on the desk of George Barker who realised that apart from a few SSSIs there was little that the NCC could provide. It certainly was not possible to provide the strategic overview for nature conservation that the County required. So Barker commissioned Bunny Teagle, who had considerable experience of London's natural history, to carry out a comprehensive survey of wildlife in the West Midlands. The results were published in 1978 as *"The Endless Village"*. If you have not read it, then you should because this was the book that made things happen. Not only does it provide a remarkable description of the natural history of the West Midlands but it also goes on to propose a philosophy and strategy for urban nature conservation that might apply anywhere. This

was in one slim A4 volume. It remains a seminal treatise on this subject that ought to be on your bookshelf.

Bob Boot, then Director General of the Nature Conservancy, saw the implications and took on Land Use Consultants to develop the NCC's ideas further. They put this in the hands of landscape architect Lyndis Cole who had recently completed a study of nature in towns and cities at Wye College and so was well versed in the subject. When I first met her she was housed down in the basement amongst the filing cabinets at the NCC HQ in Belgrave Square doing a study of opportunities for urban wildlife! As I recall virtually no one on the staff knew anything about the subject, and she had a hard time convincing people about the value of urban wildlife. However, she produced a very important report, which set the agenda for urban conservation for the next ten years or more (Cole, 1978).

Then there was Barbara Mostyn who published her important study demonstrating the personal benefits to be gained from urban wildlife projects in 1979. By the late 1970s, urban wildlife projects were underway in several cities in Britain and by 1980 urban nature conservation had reached a stage where it merited an individual chapter in the Nature Conservancy Council's Annual Report.

## Dutch initiatives

Many of the ideas at that time came from the Netherlands where initiatives in urban nature had been going on for many years. One of the key figures in the early days was Jacques P Thijsse, a teacher from the Island of Texel who had moved to teach in Amsterdam. Having been intimately engaged with nature in the Friesian islands he missed the opportunities for environmental education, and it was he who first developed ideas for *Heem Parken* which provided examples of natural habitats for city children to explore in the places where they lived. Some of these were developed in the 1930s and others in the 1950s, and together with City Farms and school nature gardens, these have become an integral part of the landscape of many Dutch cities. Another advocate was Louis Le Roi, an art teacher and gardener who argued for people to be allowed to garden on unused land, and promoted the creation of ecological corridors within urban areas. It is the kind of thing we are very familiar with now in city wildlife projects, but in the 1970s and early 1980s the Dutch were way ahead of us (Ruff, 1987). These ideas were imported into the UK by landscape designers like Alan Ruff, together with Bob Tregay who worked on the design of Warrington New Town (Ruff & Tregay, 1982). At that time in the UK it was truly innovative to be designing naturalistic woodland communities as the matrix for a housing scheme. These woodland belts have now matured and a visit today to the Oakwood and Birchwood districts of the New Town clearly demonstrates the success of this ecological approach.

## Changing attitudes

It's worth considering in a little more detail the context in which urban wildlife conservation was developing in the late 1970s and early 1980s. Yes, there was a new movement but it was fragmented and not supported by the mainstream policies of official agencies or the main voluntary sector bodies. The NCC was still firmly rooted in criteria for site preservation based on intrinsic scientific interest. The *Nature Conservation Review* (Ratcliffe, 1977) was based firmly on scientific quality, including habitat and species rarity, degree of diversity, along with intactness, and size. But the list of criteria also included intrinsic appeal. Somehow that was allowed, though it was rather ill defined. It was essentially an acceptance that some forms of nature are more appealing to the general public than others. So places that are notable for their birds, or colourful wild flowers such as orchids, might be regarded as more important than habitats of value for spiders, beetles or bryophytes. As a criterion intrinsic appeal crossed the boundary into the realm of public awareness and appreciation rather than strict science. Yet it was to be another twenty

years before promotion of public awareness of nature became a mainstream policy of Government agencies.

In the early nineteen-eighties a few places in cities were protected as SSSIs because they had certain features which were important. So Barn Elms Reservoir in west London was protected because it supported a significant wintering population of Smew, in sufficient numbers to warrant SSSI status. Similarly the Welsh Harp Reservoir in Brent was scheduled largely because it had a very large breeding population of Great Crested Grebe. Other places that were protected at that time included Perivale Wood, a tiny fragment of ancient woodland in west London owned by the Selborne Society. It was also one of the first Local Nature Reserves, despite the fact that it was only open to the public about one day a year! People, it seemed, were not important.

The county Wildlife Trusts and their umbrella organisation the RSNC gave little support to urban wildlife conservation. It simply wasn't on their agenda. Most committee members were only interested in the real gems of the British countryside, places like Wicken Fen, Blakeney Point, Cley, Spurn Head, Upper Teesdale or the Derbyshire Dales. Yes, these were important places, and they are still key areas for biodiversity. But there was no recognition of the potential value of places in towns and cities. The County Trusts had no link whatsoever with the urban scene, which led to big tensions with the rapidly developing urban wildlife groups.

Urban wildlife initiatives tended to result from the actions of local people who wished to protect something under threat. That was very often the way things started, and most of these sites were very ordinary, with nothing special or rare about them. Action was often stimulated by landscape designers working in the urban environment, rather than by ecologists or planners. Indeed in the 1970s there existed what was once described as a 'black hole' between ecology and planning, which even influenced the effectiveness of national agencies dealing with nature conservation and countryside issues.

Things started to change in the early 1980s. There are many cases where new arguments were used to fight developments as the value of nature to local people became established. Gunnersbury Triangle in Chiswick west London is a good example. It was semi-natural secondary woodland on an area that had once been allotments, but had developed naturally by colonisation since the 1930s. Arguments from the developers were based on traditional scientific criteria, and as a result they came to the conclusion that there was nothing important there. In contrast I argued on behalf of the GLC that this site was really important to local people because there was nothing like it within a reasonable distance where people could have access to nature. Fortunately the Inspector agreed that the area was important to local people and dismissed the appeal. The Triangle later became one of London Wildlife Trust's flagship sites. That was the first instance in London where these new criteria were used to argue the case, but many followed (Goode, 1986).

I well remember the disused filter beds along Lea Bridge Road in East London which had been transformed by natural colonisation into an unofficial wildlife haven. They were surrounded by a high brick wall which was broken in places and you could climb in. My first visit was early one morning and I found a veritable oasis with Kingfishers, Herons and Pochard – and as I walked round snipe took off from the marshy edges. It was an amazing place totally surrounded by industry and housing. Yet it had all the values of a mini nature reserve. When it came up for development, local people scrawled *"Save our filter beds"* on the wall. It was saved by the Lea Valley Park Authority which bought the land and added it to the park.

Moseley Bog was a 'cause celebre' in Birmingham round about 1980 when it was also threatened with development and there was a lovely story of a boy sitting up a tree

when a local government inspector came round to look at the site. The boy said, *"Are you the man who's come to save the Bog?"* That was a moment of truth for the Inspector.

## The 1980's: a decade of action

From 1980 onwards there was rapid development of urban wildlife groups, spearheaded from the West Midlands where Chris Baines was very involved, but also in towns and cities throughout the UK where there were about 80 groups by 1985. They still had an uneasy relationship with the county Wildlife Trusts, and frankly I think they still do even now in 2005. There were lots of examples of antagonism. The City Wildlife Project in Leicester was part of the Leicester and Rutland Trust for Nature Conservation, but that really upset many of its members who said, *"We're not interested in this city focus, - this is socialist, 'reds-under-the-beds' stuff."* It really was not relevant to the County Trusts. There are suggestions that these early urban wildlife groups were the equivalent in nature conservation terms of the Tribune Group in politics, i.e. those developing a more radical agenda ahead of their party, and I think there is some truth in that. Many of the ideas that came out of these urban groups, especially those concerned with people and nature, have subsequently been adopted in the wider countryside.

But some actions of the urban wildlife network during the 1980s were less helpful, as can be seen from the first issue of *Urban Wildlife* (Rowe, 1987) which records strong antipathy to establishment of a new Urban Commission proposed by the Government in 1987. Many urban wildlife groups were opposed to the role of Groundwork in urban regeneration, and were dead against the idea that it should become the urban equivalent of the Countryside Commission. Urban wildlife groups were too insular in their approach, unwilling to go beyond a relatively narrow remit, and at the same time were intensely protective of that remit. They said, *"We don't want a new Commission. We do our own thing, but we want the money"*. In retrospect I think they missed out on a huge opportunity. Only now, twenty years later, with the formation of CABE Space are we gradually recovering lost ground. [Sadly CABE Space was subsequently merged with the Design Council in 2011 and seems to have lost its ecological focus. Ed.]

What else was going on in the mid-eighties? The new metropolitan counties of London, West Midlands, Manchester, Tyne & Wear, and Merseyside introduced many innovative programmes. Lots of strategies for urban nature conservation were produced and they make fascinating reading. They had aims that set out very clearly the values to people, the importance for local communities and the vital need for protection of sites that were being identified within each of those regions. Manchester, Leicester, Sheffield, Leeds and many others had detailed schemes, but the coverage was patchy. Nevertheless it is interesting to look back to see what was going on. I am particularly familiar with the situation in London which was one of the first to develop an urban nature conservation strategy. When I was appointed as the first Senior Ecologist at the Greater London Council in 1982 we published an introduction to nature conservation for politicians which not only set the scene but also included a set of ecological policies as part of an on-going revision of the Greater London Development Plan (GLC, 1984). This document explained why nature conservation was important in a big city and it provided the basis for getting politicians onside.

I quickly realised that we urgently needed to do two other things. First we needed a complete survey of all the areas of potential wildlife value, and secondly we needed an appropriate set of criteria to decide which of these places deserved protection. As well as intrinsic scientific interest, these criteria included other factors like typical urban character, cultural and historic character, geographical position, and access for educational use. These sorts of things became increasingly important, if not more important, than some of the traditional criteria. As a result of this evaluation areas of importance were divided into three categories. Metropolitan Sites were the most important on a London-wide basis, of which there were about 130. Others were identified at Borough level, and yet others locally. The whole series amounted to a total of 1500 sites. After the abolition of the GLC in 1986 the London Ecology Unit was formed to continue this work. We published a set of handbooks identifying and describing all the important sites in each London Borough. These were then protected in each borough's Local Plan. So, although essentially a non-statutory system their protection became statutory through adoption as part of the planning process (Goode, 1999).

By the year 2000, we had published 31 handbooks that together provided an invaluable basis for conservation in London. Some of the ideas that were developed in the early days like Areas of Deficiency of Nature were further developed and refined so that now the whole of London has been mapped for such areas. These are defined as areas more than 1 km from an accessible wildlife site. These have all been mapped using real walking routes, which involves very meticulous detailed analysis. So it is possible now to consider where it might be possible to mitigate the problem by, for instance, building a footbridge over a railway line to allow deprived sectors of the population to have access to particular sites. These maps are also being used by planners in local boroughs to identify potential new areas for nature conservation as part of the London Plan.

So the series of handbooks in London set the scene for better understanding of nature in the city, took nature conservation into the planning process, and became an integral part of the *London Biodiversity Strategy*. Crucially this strategy now forms an essential part of the statutory London Plan developed by the Mayor and Greater London Authority (see Goode 2005 and GLA 2002). I have referred to these handbooks as the ecological equivalent of Pevsner.

# Creating new habitats

The interface between ecology and horticulture provided one of the most significant aspects of urban conservation through a new emphasis on ecological restoration, habitat management and habitat creation. During the 1970s I was aware that park managers in Glasgow were being encouraged to plant wild flowers in their parks and they were gaining much public support for this. Later I discovered that Grant Luscombe was one of the great pioneers in developing this kind of work through Landlife, a voluntary body working with local communities in Merseyside. Landlife still continues with great success today. Until recently the national Urban Forestry Unit [now sadly demised - Ed.] provided an invaluable research input in the field of ecological regeneration. These organisations brought considerable professionalism to the subject.

There were others too. The William Curtis Ecological Park near Tower Bridge in London, the brainchild of Max Nicholson, was developed in 1978 on a shoestring budget of a few thousand pounds. The aim was to convert a lorry park on a piece of derelict land into an ecological park, using the same ecological principles as those used to create the heem parks of the Netherlands. It was designed by Lyndis Cole, that same lady who was working in the basement of Belgrave Square, who was out there putting her ideas into practice as a landscape designer. It was a very successful project and there was a huge pioneering spirit in developing such sites at that time. Although they tried to create rather too many different habitats on such a small site the ideas were good and it was used very effectively. A total of 120,000 children used it as an educational site over 8 years, until it closed in 1986. (Ironically my recent office at the Greater London Authority looked out over the very spot that used to be the park). At the William Curtis Park, the staff were particularly good at monitoring. Detailed records were kept and these demonstrated the range of species colonising such sites. The records of butterflies and birds were particularly valuable. Over several years there was a significant increase in butterflies, including species that wouldn't normally occur in the centre of town, and this only happened because of the range of plant species, especially clovers and vetches, that Lyndis Cole had incorporated in the design. If all these early sites had been monitored to record the changes over the years we would have a much better body of ecological information.

There were other ecology parks during the 1980s including Walker Gate and Benwell Nature Park in Newcastle, Brandon Hill in Bristol, and Camley Street Natural Park at Kings Cross in London. The latter was developed by the GLC and handed over to the London Wildlife Trust where it became one of their most important sites. It has survived the threat of destruction from the Channel Tunnel Rail Terminal and remains an invaluable resource to the local community.

We were not alone in Britain developing such sites. Ideas spread rapidly. One particular milestone in the mid-1980s was Toronto Ecology Park, together with its Ecology House, a sustainable building, with a host of innovative features, even organic loos. Sadly the features that were demonstrated then are still not happening in any major way today. The John Inskeep Environmental Learning Centre in Portland, Oregon was particularly interesting. I was there in 1986 and was fascinated by the fact that it combined a centre for nature conservation together with recycling. In the middle of a mini nature reserve, reclaimed from a university car park, was a recycled building and all the paths and walkways were made of recycled plastic not wood. The environmental centre itself was constructed from timber reclaimed from other buildings, so the idea was *"We don't waste, we recycle and reuse"*.

# Broadening the agenda

The development of that whole approach required a new philosophy and new techniques involving collaboration between the disciplines. One of the key meetings in the

1980s was on ecological landscape design. This was a joint meeting between the British Ecological Society and the Landscape Institute which provided new ideas within the landscape profession and the proceedings turned out to be one of the best-selling books the BES have ever produced (Bradshaw *et al.*, 1986). At about the same time an international meeting on urban ecology was held in Berlin, which for the first time focussed the attention of professional ecologists on urban habitats and species (Bornkamm *et al.*, 1982). It is time we had another meeting of this kind.

Then in 1984, came an important meeting in Liverpool, the joint UK/US meeting linked to the Liverpool Garden Festival. This was a really pivotal meeting for a number of reasons. It brought together a lot of people from all the professions, from landscape design, park management, ecology and urban regeneration. The event also involved practitioners from the US. People talked a lot about Olmstead and his legacy of urban parks, but others were more concerned with the problems of the Toxteth and Brixton riots of 1981. The need to engage all the professions in urban regeneration was at the top of the political agenda. That was clearly uppermost in the mind of Michael Heseltine, then Secretary of State for Environment, when he came to the meeting and announced the formation of a new UK Groundwork Foundation. The Groundwork Foundation was built on work that that had been done so successfully by John Handley in St Helen's and it went on to become one of the leading players over the next twenty years.

This was followed in 1987 by a meeting in Toronto. "Greening the City" was a reciprocal meeting with people from the Liverpool congress, but it had a much wider sustainability agenda. In effect, it was posing the question, "*What are the ecological issues and how do we find saner ways of urban living?*" This paved the way for alternative thinking on sustainability in a much wider context, addressing the problems of Mexico City and Calcutta alongside those of western cities. It was the first attempt to address this

wider agenda of urban management and ecological issues, and it didn't sit easily with the more constrained area of urban wildlife conservation. However, it was an important meeting that demonstrated the need for interdisciplinary approaches (Gordon, 1990). Meanwhile the National Institute for Urban Wildlife in the States organised a meeting in 1986 in Washington DC (Adams & Leedy, 1987), which brought together urban planners, landscape designers, and researchers, as well as practitioners dealing with urban wildlife management. This was another important meeting that raised some fundamental questions about the newly developing philosophy of urban nature conservation. A striking feature of this conference for a visitor from the UK was the large number of delegates of black and ethnic minority backgrounds working in the environmental professions.

# Involving local people

Involvement of local people was an emerging and distinctive theme that set urban nature conservation apart from more traditional approaches. It led eventually to a rethink of official government policy, with recognition that nature matters to people in towns and cities and that catering for public awareness is just as important in the wider countryside. It took until 2002 for these ideas to gain acceptance, such is the inertia of the official system. However, despite much of the work on social values by Barbara Mostyn, Carolyn Harrison & Jacquelin Burgess, this is still the least developed and least understood area of urban nature conservation (Harrison *et al.*, 1987, and Millward & Mostyn, 1989). There is room for much more work in this field which crosses the boundaries between the ecological and social sciences. A report on the East Midlands in 2003 raised more problems than it answered and I think there are still a lot of issues that need to be investigated.

# Resume

To summarise: by the end of the 1980s there was a new philosophy valuing common and familiar species and making nature accessible to people in towns and cities. This philosophy accepted that man-made habitats can be of great value, and promoted the enjoyment of nature. For example one of the policies in the Greater Manchester Strategy, was to *"Ensure that the enjoyment of nature is available to all."* Policies and programmes were adopted by official agencies to protect areas of value and to create new habitats in areas of deficiency. Other significant advances were the emergence of ecological landscape design as a significant element within the landscape profession, and acceptance of habitat creation by ecologists working with other professions to seek opportunities for nature within the urban fabric. Finally there was spontaneous development of a bottom-up approach involving local groups in community schemes which were widely supported to encourage local ownership of these initiatives. This was in marked contrast to the top-down official approach which had alienated many local people. For more details see my paper in the *'Journal of Applied Ecology'* (Goode, 1989) which reviewed progress up to the late 1980s and the final chapter of Wild in London (1986) entitled A Green Renaissance. I have to say, that the decade of the 1980s was a most rewarding period for those of us working in urban nature conservation. It seemed at times that almost anything was possible.

Although much progress had been made there was still a problem. Urban nature conservation was still far from the mainstream; I suspect it still is. It was not integrated into urban development and was not on the agenda of the key professions, the architects, the civil engineers, the house builders and developers, often not even park managers. There was still a lack of skills for many aspects of habitat creation and a lack of awareness of the opportunities on the part of urban designers. This was despite the publication of a plethora of books on the subject. These included a practical guide from the Ecological Parks Trust on *Promoting nature in cities and towns* (Emery, 1986); a wide ranging and compelling philosophy by David Nicholson-Lord (1987); Ann Spirn's *Granite Garden*, a landscape designer's view of cities as ecosystems (1984); Oliver Gilbert's seminal work of 1989 which remains the best available treatise on urban ecology; *Green Cities*, a summary of the Toronto meeting by David Gordon in 1990; Jacklyn Johnston's ground-breaking book on *Nature areas for city people* (1989) and several books specifically about urban wildlife including Baines (1986) and Goode (1986). We were not lacking information.

# Looking to the future

So, twenty years later, where are we now? In many respects the philosophy remains the same. The difficulty has been implementing it. Most towns and cities now have policies for the protection of urban wildlife sites, which is achieved largely through the designation of Sites of Importance for Nature Conservation (SINCs) by local planning authorities. Some of these programmes have been remarkably successful in ensuring the protection of sites through the planning process, despite the absence of a statutory basis for such procedures. Some cities have been better at it than others. The strategies for Bristol and Sheffield are particularly good examples.

Perhaps the most successful schemes are those which have been integrated firmly into local government arrangements, through voluntary joint committees representing district planning authorities, as in the case of London and Manchester. These have developed clear strategies for protection of important sites through the planning process. The London example is unusual in that there is now a statutory requirement for a Biodiversity Strategy. It is perhaps ironic that it is the only part of the UK where such a statutory duty exists. But it demonstrates what is possible (GLA, 2003).

In all these cases the recently published Defra guidance on identification of Local Sites helps enormously (Defra, 2006). This

Guidance builds on existing good practice and provides the basis for selection and protection of Local Sites, recognising their importance as a component of the national series of biodiversity sites. It gives recognition to the value of such sites in urban areas stating, "*In populous areas that are poorer in high quality natural environment, sites of lesser intrinsic ecological interest may still be of substantive nature conservation value for the opportunities they provide for the appreciation of nature*". It also specifically recognises the value of such sites for environmental education. The Defra guidance goes a long way to formalising what has already become accepted practice in many towns and cities where Local Authorities have taken the lead in developing programmes for the protection of such sites.

Local Biodiversity Action Plans have also helped to put ecological issues on the urban agenda. Their partnerships have helped to broaden the constituency involved, with greater integration between the professions, and with local people. But such Action Plans need to be made mandatory for all towns and cities. Whilst local action plans have helped, the national Biodiversity Strategy has failed to deliver on urban conservation. Although this aims to provide targets for the full range of habitats occurring in the UK, a priority habitat has yet to be identified for the urban environment. The absence of a designated priority habitat was highlighted in a report for English Nature (Tucker *et al.*, 2005). A post-industrial habitat of open land has been proposed, which if implemented would go some way towards conserving the range of specialist species characteristic of urban commons (Gilbert, 1992).

The role of biodiversity in providing ecosystem services is perhaps the most promising area for future development. Too often, biodiversity is the last thing to be addressed in sustainable development strategies. Yet it affords considerable opportunities for improvements in urban design to cope with climate change. We still have a long way to go before this is addressed as a matter of course in urban planning, but I am encouraged by proposals made by the Royal Commission on Environmental Pollution which has argued for greater use of the green infrastructure as a basis for environmental sustainability. This would include use of green corridors and multifunctional greenspace as suggested by Barker (1997), but also includes greater emphasis on biodiversity within the built environment, such as green roofs and Sustainable Urban Drainage Systems. We lag behind several European countries in our use of these approaches which could significantly improve the environmental performance of urban areas. The evidence is not lacking. The London Ecology Unit published guidance on green roofs (Johnston and Newton, 1993) and English Nature provided an excellent guide to the benefits of green roofs in 2002. There is similar material available for SUDS (CIRIA, 2001). What is now needed urgently is the will to change our practices as part of our adaptation to climate change. The TCPA published an excellent design guide for biodiversity which, if implemented, would bring urban planning into the 21st Century (TCPA, 2004). The London Development Agency also published guidance on planning for biodiversity (LDA, 2004). Other regional development agencies need to follow suit.

What else has happened? Lottery funding has had a huge impact not only by procuring important habitats for protection, but also providing funds for the development of numerous high quality visitor centres. The London Wetland Centre is just one example. The whole agenda dealing with people and wildlife in urban areas is more formalised, and forms an important part of national strategies for conservation. For instance, the need for accessible natural greenspace within urban areas has been adopted as one of the key priorities for Natural England, taking the standards developed by Box & Harrison (1993), Harrison *et al.* (1995) and Handley *et al.* (2003), and adopting a target of 300m as the desired distance from home to an accessible wildlife site. At the same time CABE Space has recognised the importance of biodiversity in urban greenspace, publishing detailed guidance on "making contracts work for wildlife" and encouraging greater public use (CABE Space, 2006). These various initiatives were described in more detail in a chapter on nature conservation in towns and cities (Goode, 2007) and a report on Green Infrastructure commissioned by the Royal Commission on Environmental Pollution (Goode, 2006).

What else has happened? Lottery funding has had a huge impact not only by procuring important habitats for protection, but also providing funds for the development of numerous high quality visitor centres. The London Wetland Centre is just one example. The whole agenda dealing with people and wildlife in urban areas is more formalised, and forms an important part of national strategies for conservation. For instance, the need for accessible natural greenspace within urban areas has been adopted as one of the key priorities for Natural England, taking the standards developed by Box & Harrison (1993), Harrison *et al.* (1995) and Handley *et al.* (2003), and adopting a target of 300m as the desired distance from home to an accessible wildlife site. At the same time CABE Space has recognised the importance of biodiversity in urban greenspace, publishing detailed guidance on "making contracts work for wildlife" and encouraging greater public use (CABE Space, 2006). These various initiatives were described in more detail in a chapter on nature conservation in towns and cities (Goode, 2007) and a report on Green Infrastructure commissioned by the Royal Commission on Environmental Pollution (Goode, 2006).

However, there is still a need for a critical analysis of the economic and social value of nature and I believe this to be crucially important in the context of urban development, especially as it relates to the brownfield debate. Such an analysis needs to consider the potential for biodiversity in the context of the compact city, taking into account people's needs for health and wellbeing. These arguments have been well made (Bird, 2004 and Douglas, 2005) but they

have yet to be widely accepted within the economics of urban development. [Some progress was made since this lecture was given, but the changes since the 'new austerity' was imposed have set progress back considerably – Ed.]

Too often biodiversity is the last thing to be addressed in sustainable development strategies; even today this still applies. We will have a long way to go before the vision of an ecologically sustainable city comes to fruition. This remains the greatest challenge for urban nature conservation, making the case for nature in economic terms that will stand the test of time. This will require a new philosophy based on people's needs which will become all too apparent when climate change begins to have real impacts, and the economic and social values of biodiversity become too great to ignore. We shall then need another decade of action like the 1980's.

[Today, even more than when David gave his lecture, the challenges remain great. Now, with expanding cities and urban areas, combined with cuts to local authority and agency conservation and countryside services, the need for increased action rather than less, remains paramount – Ed.]

Author's note: Much of the material contained in this paper was subsequently incorporated in chapters on Urban Nature Conservation in the New Naturalist volume *Nature in Towns and Cities*, 2014, Harper Collins.

# References

Adams, L.W. & Leedy, D.L. (Eds.) (1987) *Integrating man and nature in the metropolitan environment*. Proc. Nat. Conf. on Urban Wildlife. Nat. Inst. for Urban Wildlife. Columbia, Maryland, USA.

Baines, C. (1986) *The Wild Side of Town*. BBC and Elm Tree Books, London.

Barker, G. (1997) *A framework for the future: green networks with multiple uses in and around towns and cities*. English Nature

Research Report **256**, Peterborough.

Bird, W. (2004) *Natural Fit: Can Green Space and Biodiversity Increase Levels of Physical Activity?* Report for the Royal Society for the Protection of Birds.

Bornkamm, R., Lee, J.A. & Seaward, M.R.D. (1982) *Urban Ecology*. The 2nd European Ecological Symposium. Blackwell Scientific Publications, Oxford.

Box, J. & Harrison, C. (1993) Natural Spaces in Urban Places. *Town and Country Planning* **62** (9) 231-235.

Bradshaw, A.D., Goode, D.A. & Thorpe, E.H.P. (Eds.) (1986) *Ecology and Design in Landscape*. Symposium of the British Ecological Society, 24. Blackwell Scientific Publications, Oxford.

CABE Space (2006) *Making contracts work for wildlife: how to encourage biodiversity in urban parks*. Commission for Architecture and the Built Environment, London.

CIRIA (2001) *Sustainable urban drainage systems: Best practice manual.* CIRIA Report C523. CIRIA, London.

Cole, L. (1978) *Nature Conservation in Urban Areas*. Nature Conservancy Council. London

Davis, B.N.K. (1976) *Wildlife, urbanisation and industry. Biological Conservation*, **10**, 249-291.

Defra. (2006) *Local Sites: Guidance on their Identification, Selection and Management.* Defra, London.

Douglas, I. (2005) *Urban greenspace and mental health.* Discussion paper of the UK MAB Urban Forum.

Emery, M. (1986) *Promoting nature in cities and towns: a practical guide.* Ecological Parks Trust. Croom Helm. London.

English Nature. (2003) *Green Roofs: their existing status and potential for conserving biodiversity in urban areas.* English Nature Research Reports. Number **498**. English Nature, Peterborough.

Fairbrother, N. (1970) *New Lives, New Landscapes.* Architectural Press, London.

Fitter, R. (1945) *London's Natural History.* Collins, London.

Gilbert, O. (1989) *The Ecology of Urban Habitats.* Chapman and Hall, London.

Gilbert, O. (1992) *The flowering of cities: The natural flora of urban commons.* English Nature, Peterborough.

GLA (2002) *Connecting with London's nature: The Mayor's Biodiversity Strategy.* Greater London Authority, London.

Goode, D. (1986) *Wild in London.* Michael Joseph, London.

Goode, D. A. (1989) *Urban nature conservation in Britain.* J. Appl. Ecol. **26**, 859-873.

Goode, D. A. (1999) *Habitat survey and evaluation for nature conservation in London.* Deinsea 5: 27-40. Natural History Museum of Rotterdam.

Goode, D. (2005) *Connecting with nature in a capital city: The London Biodiversity Strategy.* In: T. Trzyna (ed.). *The Urban Imperative.* California Institute of Public Affairs, Sacramento, California for IUCN. See www. iucn-urban.org

Goode, D. (2006) *Green Infrastructure.* Report commissioned by the Royal Commission on Environmental Pollution. www.rcep.org.uk

Goode, D. (2007) Nature conservation in towns and cities. In: *Contemporary Rural Geographies: essays in honour of Richard Munton.* Edited by H. Clout. Routledge

Gordon, D. (2000) *Green Cities: Ecologically sound approaches to urban space.* Black Rose Books, Montreal.

Greater London Council. (1984) *Ecology and Nature Conservation in London.* Ecology Handbook No. 1. Greater London Council, London.

Handley, J. *et al.* (2003) *Accessible Natural Green Space Standards in Towns and Cities: A Review and Toolkit for their implementation.* English Nature Research Reports. Number **526**. English Nature, Peterborough.

Harrison, C. *et al.* (1995) *Accessible natural greenspace in towns and cities: A review of appropriate size and distance criteria.* English Nature Research Reports Number **153**. Peterborough.

Harrison, C., Limb, M. & Burgess, J. (1987) Nature in the city – popular values for a living world. *Journal of Environmental Management* **25**: 347-362.

Johnston, J. (1990) *Nature areas for City People: A guide to the successful establishment of community wildlife sites.* Ecology Handbook No. **14**. London Ecology Unit, London.

Johnston, J. & Newton, J. (1993) *Building Green: A guide to using plants on roofs, walls and pavements.* London Ecology Unit. London. (Reprinted 2004 by GLA).

Kieran, J. (1959) *The Natural History of New York City.* Houghton Mifflin Company, Boston.

Laurie, I.C. (ed.) (1975) *Nature in Cities.* Report of the Proceedings of a symposium held at the University of Manchester. Landscape Research Group.

L.D.A. (2004) *Design for Biodiversity: A guidance document for development in London.* London Development Agency. London.

Mabey, R. (1973) *The Unofficial Countryside.* Collins, London.

McHarg, I. (1969) *Design with Nature.*

Doubleday Natural History Press.

Millward, A. & Mostyn, B. (1989) *People and nature in cities: The social aspects of planning and managing natural parks in urban areas.* Urban Wildlife Now, No. **2**. Nature Conservancy Council, Peterborough.

Mostyn, B.J. (1979) *Personal benefits and satisfactions derived from participating in urban wildlife projects.* Social and Community Planning Research. NCC. London.

Nicholson-Lord, D. (1987) *The Greening of the Cities.* Routledge and Kegan Paul. London.

Ratcliffe, D.A. (1977) *The Nature Conservation Review.* Nature Conservancy Council, London. (see Volume one, pages 6-10).

Rowe, F. (1987) News and Views. *Urban Wildlife*, 1, 2-4. Chichester.

Ruff, A. and Tregay, R.J. (Eds) (1982) *An Ecological approach to urban landscape design.* Dept. Town and Country Planning, University of Manchester.
Ruff, A. (1987) *Holland and the ecological landscapes 1973-1987.* Urban and Regional Studies. Delftse Universitaire Pers.

Spirn, A.W. (1984) *The Granite Garden: urban nature and human design.* Basic Books Inc., New York.

Teagle, W.G. (1978) *The Endless Village.* Nature Conservancy Council. West Midlands Region.

T.C.P.A. (2004) *Biodiversity by Design. A guide for sustainable communities.* Town and Country Planning Association, London.

Tucker, G, Ash, H., & Plant, C. (2005) *Review of the coverage of urban habitats and species within the UK Biodiversity Action Plan.* English Nature Research Report Number **651**, Peterborough.

# *Wastes and strays: the archaeology of urban commons*

**Mark Bowden**

English Heritage, Senior Archaeological Investigator

## Abstract

This was an English Heritage national archaeological project, in the form of an exemplary study (through non-intrusive survey) of selected urban commons.

## Need and Purpose

Urban commons are an almost unexplored archaeological resource. Nevertheless, they offer the opportunity to extend our attention within the historic urban agenda to a wider range of archaeological phenomena and to a broader topography beyond the built-up core that has generally been the focus of study. These commons can also offer a new window of understanding into the development of the adjacent town or city. This is a matter of recognising what has previously been 'not known because not looked for'.

Just as importantly, such spaces are critically important in the modern urban fabric. For the inhabitants of a town they are key components in what they would understand as their 'historic environment' as it impacts on their daily lives. Urban commons are almost chronically under pressure, both from major developments and from more insidious and cumulative causes. They, typically, had no Conservation Plan. Their detailed Historic Environment content and its value is certainly unknown and therefore delivers no conservation benefits.

EH had current initiatives within its urban research framework addressed at urban parks and cemeteries. This project directed attention to another sort of urban space. It sought to establish ways of placing Historic Environment value on locations that are undoubtedly publicly valued. It also seeks to highlight, by example, the recognition and use of those HE values, both by local authorities and the public. In those terms it responds to the aspirations set out in *Power of Place* and is a contribution to the agenda expressed in the paper on *Urban Priorities 2002–05*.

## Aim and Objectives

### Aim of the project

To investigate the archaeological content and Historic Environment value of urban commons in England and to prompt appropriate conservation strategies for them.

### Objectives

i. To research and survey a representative sample of urban commons in England.
ii. On that basis, to produce a synthetic account of urban commons, highlighting both historical and conservation aspects.
iii. To make available the results of that work in the most appropriate ways to the widest constituency.
iv. Through partnership arrangements in the individual sample places, to promote and underpin local community conservation initiatives.

### Relevance to other initiatives

The project provided a direct response to a request for survey work on the Lincoln commons, as a next step to the Urban Archaeology Database / Research Strategy work there by EH Programmes Division. It also seeks to extend and generalise from that. It should therefore have provided a lead and example to other UADs and be relevant to many lesser towns within Extended Urban Survey projects.

The work provided additional context to the work of the EH Urban Panel. It gave EH regional staff a basis for conservation initiatives in selected instances and provided a framework for casework elsewhere.

The study prompted selected local authorities to promote conservation and to support that action through partnership in the project's initiatives. It was anticipated that each individual surveys should, where possible, entail a partnership of some kind with the local authority would ensure delivery of conservation benefits (e.g. a Conservation Plan) and public information gains (e.g. community leaflets, etc.)

It was to supply the National Monuments Record with new information and upgrade the relevant local authority UADs.

The activities were the basis for exploring which other agencies, alongside EH, had a common interest in the conservation issues of these landscapes. In particular, it sought to open up opportunities for complimentary botanical and biodiversity studies.

## Outcomes

Outcomes were intended to include:

i.  Enlarged sense of the urban fabric.
ii. Identification of the EH component of urban commons and articulation of its value.
iii. Support for the UAD programme – and for the sense that it is a beginning, not an end.
iv. Promotion of conservation initiatives about urban commons.
v.  Promotion of public appreciation of urban commons and their historical value.
vi. Raised academic awareness and new research agendas.

*International Urban Ecology Review*, 5, Urban Environments Issue

# Nature in your neighbourhood: helping local communities improve green spaces for wildlife

**Rachel Remnant**
Cambridgeshire & Peterborough Biodiversity Partnership

## Introduction

*Nature In Your Neighbourhood* (NIYN) was a programme created and run by the Cambridgeshire and Peterborough Biodiversity Partnership. The project put the 'action' into Urban Biodiversity Action Plans by helping local communities to make environmental improvements to green spaces and neglected pockets of land. Sites were chosen for their wildlife interest and potential, and for a community commitment to turn the areas into a local resource. Local groups could then tap into a network of expertise to help them achieve their goals. The Partnership was made up of Cambridgeshire County Council, the district councils, the Wildlife Trust, the RSPB, Anglian Water, English Nature, the Environment Agency and Landscape 2000.

By the early 2000s, twenty-two local community groups were helped by the Partnership. They had planted grassland, woodland and wildflower meadows, coppiced trees to let in more light for the ground flora, installed bird and bat boxes and encouraged healthy eating with fruit and herb planting. It had previously been difficult for groups to access the relevant resources needed to carry out environmental enhancement work and to coordinate their efforts. The NIYN project provided a network of like-minded local groups and the emphasis of the scheme was to form sustainable projects that would continue to thrive after initial help had been given.

The project had enabled community groups to get access to the resources and expertise of the partnership organisations and had empowered them to take positive action to improve their environment for both wildlife and people. This had proved very popular and volunteers commented on how useful it was to have a point of contact, through the network,

that helped them to establish their project, keep the momentum going and bring their ideas to fruition. The work encouraged wildlife and brought other environmental gains, and increased community spirit. It also helped to improve health and well-being. Taking a role in improving the local environment gave communities a sense achievement and of ownership over their neighbourhoods.

## Case Studies

### Harston Recreation Ground

**Harston Recreation Ground** was referred from the County's Countryside Services Team. Residents were complaining about the lack of light in their gardens due to trees in the recreation ground. A coppicing day was held to reduce this problem and provide a mosaic of habitats in the tree belt along the edge of the recreation ground. The village has taken to NIYN like a player of dominoes: first there was the coppicing day, then the adjacent apple and damson orchard came in and more recently the burial ground has had a management plan written to incorporate biodiversity measures. Although only two village residents organised what's happening, but they always had a good turn-out to their work parties or apple day. This was where NIYN was most helpful – supporting a minority who would otherwise be left with too much to do.

### Edgecombe Flats Green

**Edgecombe Flats Green** was located in Kings Hedges, Cambridge, a modern suburb of the city, with young green spaces. The City Council planning department planted swathes of native trees in 1992. When NIYN began they were keen for the project to encourage more interest in the site as a nature space. For the past two years NIYN had secured funding from the landowners, City Homes, to carry out wildflower planting with local residents. Community involvement was not very high, just two or three stalwart volunteers. A new approach was being trialled – for example, taking local nursery school children on 'discovery walks'. Although local interest is been low, the City Council hopes that once the trees are older and larger the impact of the space will be felt more significantly, which will generate more interest in biodiversity. The consolation is that while the green is underused, it is not disturbed!

## Lolworth Wildflower Meadow

**Lolworth Wildflower Meadow** was a glorious corner of an idyllic village. With no through road, children are free to play on the streets and are entertained off road on Fridays with their 'Kids Club'. NIYN first encountered the meadow on one such afternoon, when Rob Mungovan, district ecology officer led a wildflower walk around their old allotments. Only one true allotment remained, the rest had fallen into disuse, so a local farmer had bought local seed to sow after turning the soil with tractor mounted machinery. Once the blooms were over, there was a return to true horse-power; the meadow was cut by a horse drawn mower. The village takes great pride in its community spirit and raked up the hay within an hour, it is easy to assemble a large group of volunteers in Lolworth!

*International Urban Ecology Review*, 5, Urban Environments Issue

# *Recombinant communities along an urban routeway*

## Hilary Thomas
University of Wales College, Newport

## Abstract

This project sought to join history, biodiversity, and cultural influences as they affected a major highway development in South Wales. The road project studied was the Barry Docks Link Road in South Wales.

## History

The background to the study area included the:

- **Reason for initial construction;**
- **Surrounding land use;**
- **Soil parent material;**
- **Time zero...1980;**
- **Construction and initial management.**

## Impacts

The Link Road development affected the Northern woodland, agriculture in the Vale of Glamorgan, and was associated with the outward spread of the suburbia of the town of Barry.

One initial result of the project was landscape simplification and associated impacts on biodiversity.

- **Fluctuating species numbers;**
- **The effects of increased environmental stress;**
- **History as a continuing driving force toward diversity;**
- **Some effects were unexpected and unpredicted...**

## Culture

As the ecology developed it was clear that there were strong local influences and inputs of species from the local landscape.

Success of colonisation was affected and influenced by a variety of other factors.

## Conclusions: a recombinant future?

The story so far ...

- **Top soiled areas always far more species-rich;**
- **An increasingly stable grass-based cover upon the less hostile natural parent materials;**
- **Input from surrounding land cover.**

Over time since establishment, the trend has been towards a 'dynamic equilibrium' of native and other species, many colonising in from local landscapes and ecosystems. These trends were driven by local factors and key environment influences.

Observations over time demonstrated the influences of environment and of people in the emerging ecologies along the new road.

**History > Biodiversity < Culture**

# A methodology for enabling companies to establish and implement biodiversity action plans within environmental management systems

**James Calow, MIEEM, AIEMA, AMIEnvSc**
Environmental Systems Research Group, School of Engineering & Applied Science, Aston University, Aston Triangle, Birmingham, B4 7ET

## Abstract

Due to improved legislation covering species and natural habitats, and gradual improvement in the consideration of ecology under Environmental Impact Assessment, development projects in the UK are showing greater awareness of biodiversity impact and mitigation. However, operational business management takes little account of the natural world, thousands of companies have so far been accredited to ISO 14001 and yet there have been very few examples of biodiversity issues being formally linked to certified Environmental Management Systems (EMS).

As a first step in conserving biodiversity issues in a commercial setting, there is a need to develop structured guidance setting out the actions companies must take to formally incorporate biodiversity issues into their Environmental Management Systems. Between October 2000 and October 2003 a research project was undertaken at the Environmental Systems Research Group at Aston University to develop a methodology to enable a company to establish and implement a corporate Biodiversity Action Plan within an Environmental Management System.

This paper presented an overview of the research project with specific emphasis on a framework for incorporating corporate biodiversity information into an environmental management programme. In addition it described a process to enable companies to identify and quantify their impacts on biodiversity, allowing them to protect and enhance their corporate biodiversity within a cycle of continuous improvement.

# Biodiversity in the urban community: the importance of involving local people in wildlife projects

## Shona Turnbull

Formerly: Hull Biodiversity Partnership,
Floor 2, Kingston House, Bond Street, Hull, HU1 3ER

## Abstract

Involving local people in urban biodiversity can be seen as a hindrance, but it would be unwise to exclude them from wildlife projects in their area. From national, regional to local level, there are a number of policies and initiatives that require the public to be kept informed and involved. Local Biodiversity Action Plans are one key way in which communities can take ownership of their urban natural resources.

Although getting the communities help can present a wide range of problems, from health and safety issues, lack of suitable training, apathy and sustainability, there are numerous benefits that outweigh these problems. Many of the solutions are cost-effective; such as using existing partner resources, press releases, fact sheets and community events. Ensure the community is on your side from the start of a new project, if you want it to succeed, or it is likely to be vandalized or at least unsupported.

## Introduction

Ecological assessments and monitoring programmes often rely on indicators to evaluate environmental conditions. These indicators are frequently developed by scientists and are expressed in technical language (Schiller et al., 2001). However, academics and scientists can sometimes lose sight of the importance of involving local people in these wildlife-based projects. These local people may be seen as a hindrance to getting on with the real work of managing a site according to scientific models of best practice. Experts tend to think that scientific facts are convincing in themselves (van Boven and Hesselink, 2002) and therefore need no explanation. However, the data have to be translated into concepts and messages that appeal to the target audience (ibid, 2002). Thus, when embarking upon a new urban nature conservation project, it would be unwise to bulldoze in, both literally and figuratively, without consulting and/or involving local people. They have to feel that the project, at least in part, belongs to them in order that they can both help support and possibly police it.

This paper examines the costs and benefits of community involvement in urban biodiversity. The emphasis will be on the Local Biodiversity Action Plan (LBAP) perspective, although many of the ideas can be applied to most urban wildlife projects.

## Methodology

The approach taken in the study was one of action research with stakeholder participants and an overview of national policy issues and context.

## Discussion of the issues and findings

### National Priorities

The government is committed to involving communities in biodiversity and has set a number of targets at national level. For example, Local Agenda 21 is about people taking action in their local environments, for the benefit of themselves, wildlife and the planet as a whole (Jones and Talbot, 1995.) This is also reflected at country level; the England Biodiversity Group recognises that people should 'be more knowledgeable about biodiversity so that they can both appreciate it and act to safeguard it' (Defra, 2003). They are also encouraging scientists to become more involved in sharing their knowledge with local communities (*ibid*, 2003).

However, the 2001 Survey of Public Attitudes to Quality of Life and the Environment showed that only about half were concerned with the loss of plants and animals in the UK (Defra, 2002), although it is a slight increase on the 1996/97 figures (Defra, 2000). Although 80% stated that they enjoyed the countryside (Defra, 2002), most people will have had more contact with wildlife in their own (urban) gardens and parks.

## Local People – Why involve them?

Eighty-nine percent of the UK population live in town or cities on 7.7% of the land (Greenhalgh & Warpole, 1995). This means that local community involvement in urban biodiversity is vital for a multitude of reasons. First of all, it is their backyard; people are most likely to access wildlife that is nearest to them. Especially in urban areas, this tends to be their own gardens, local parks or corridors used as pedestrian short cuts.

They are also an excellent source of local knowledge and expertise. This knowledge could save thousands of pounds on what otherwise may be wasted effort, as they are likely to know the history of a site. It is also often necessary for funding bids to show that there is community demand and support for urban wildlife projects.

Involving local people also helps address a number of cross-cutting themes, e.g. issues of social exclusion and crime reduction: they can be involved in management and policing of a project. Work is often carried out in areas of public access, as many biodiversity projects can be on council land, which they have access to. As you do not want your work vandalised, the public can act as informal guardians if they feel a sense of ownership, gained from being involved from the outset.

Local communities can now get a range of free and very useful information via the internet and are therefore often well informed about both wildlife issues and the threats facing them. They are therefore less likely to be fobbed off with bland excuses as to why a particular project is being handled in a particular way, especially if they feel the action is unsuitable for their area.

## Problems with involving the community

The first major hurdles to overcome when trying to involve the community in biodiversity projects are language and perception. A survey in 2001 demonstrated that only 26% had heard of the term 'biodiversity', although this is a slight increase from 22% in 1996 (Defra, 2003a). For those that do know the term, there is the added problem that biodiversity and urban is often seen as an oxymoron. We need to overcome the perception that biodiversity is something exotic or something for the countryside only. This can be done by careful management of the language used to promote biodiversity and by various educational initiatives aimed at different sections of the community.

There are a myriad of health and safety issues that must also be considered when community participation takes place. For example, if you were clearing up a derelict site, there are a range of implications that need to be considered such as needles, other sharps, leptospirosis, trip hazards and appropriate equipment handling. A full risk assessment has to be done and appropriate insurance cover provided to address these issues.

Apathy can also be an issue; LBAPs in particular are often mistakenly seen as just another part of the council, so there may not be much enthusiasm to get involved. Those that do want to participate may lack suitable skills; therefore you have to spend time and money training people to a suitable standard.

Funding can be one of the biggest hurdles to overcome. Getting money to get community wildlife projects in a city off the ground can be very difficult, due partly to the problem mentioned earlier; that of perception. Some of the larger funding organisations still tend to see biodiversity as a countryside issue, although attitudes are slowly changing. Funding is usually for a finite period, usually 1–3 years, which leads to another problem: sustainability. Wildlife projects need long-term management. After the initial public interest has waned, it often falls to over-stretched local authorities to manage a site. Changes of staff can also have a big impact. The public get quickly disillusioned if their hopes have been built up by a capable, conscientious officer, only for the momentum to be lost when funding for them or the project dries up.

When trying to get a new biodiversity project started, there can be an element of preaching to the converted. You tend to get the same people involved in lots of projects and thus they get stretched too far. Community groups are also often already committed to lots of other projects so may not have the time to commit to something new.

## Solutions to help involve the community

There are many ways that you can get local people involved, without spending lots of money. One of the simplest is to get the community involved in a 'one off' event, such as a tree planting day, so they can feel that they own a project. This can give them the psychological benefit of feeling that they have 'done their bit'.

It then allows you to get on with the rest of the project, with occasional additional help as and when needed. It is important when organising these events to keep the jargon used to a minimum when communicating with the public. For example, call projects 'wildlife events' rather than 'biodiversity action days'. It must also be borne in mind, depending on the area you are covering, that for some participants, English may not be their first language.

Table 1 summarises some of the cost-effective means that can be used to involve the local urban community in biodiversity projects. In addition, there are many other ways to gain support and help for urban wildlife initiatives. For example, in an LBAP, it is useful to use some species that are easy to identify and will have public appeal or local distinctiveness, e.g. hedgehogs, cowslips. This allows people to participate easily and leads them gently into the LBAP or other wildlife initiative processes. It is also essential to make good use of your existing partners; get them to play their part in spreading the word about particular events or species, or by providing materials e.g. leaflet production. Most urban areas have some system of small grants e.g. Lord Mayor's Trust, Community Investment Funds, Ward Forums, Rotary, local businesses or in kind donations, that can all help boost funds.

'Friends of' groups are an excellent way of encouraging community involvement.

## Table 1: Cost-effective projects aimed at community involvement in urban biodiversity

| Event/Project | Target Audience (Yrs) | Comments |
|---|---|---|
| Bug hunts/pond dipping | 0–10 | Teaches young children to take an interest in the environment |
| Wildlife gardening network | All ages | Allows people to share ideas, expertise or even plants to build up a network of biologically diverse gardens |
| Produce fact sheets/regular newsletters | All ages | Good source of participatory information/contact details |
| Wildlife surveys | Dependant on survey type | Make use of national projects e.g. natural history, Museum's woodlice survey; RSPB's garden bird counts |
| Talks/presentations | 20–80 | Community groups are often keen to become involved once they know what you are trying to acheive |

They are fairly low cost to set up and run and once they are established, they should become self-sufficient.

They can also take on some of the tasks that you would like carried out but do not have the time for yourself e.g. newsletters, publicity, wildlife video, attitudes / species surveys.

It is useful to make use of national or even international events, as they often provide free publicity materials and information. For example, International Biodiversity Day is a good time to showcase your work. Stalls at shopping centres are a good way of meeting a wide range of age groups and interests. It can also be a useful way of gathering informal data on garden species, which can be followed up by a questionnaire or survey.

At the events detailed above, members of the public have the opportunity to share their experiences of wildlife in their city. It also gives you the chance to convert them to some of your ideas, e.g. explaining the importance of so-called 'neglected' or 'scruffy' areas, which are in fact developing wildflower meadows or tree understory.

# Nature reserves, orchids and fly ash

**Peter Shaw**

University of Surrey, Roehampton

## Abstract

This paper reported on the natural colonisation of the industrial pulverised fuel ash (PFA) by vegetation, with particular reference to the appearance, explosive colonisation and subsequent decline of marsh orchids in the genus *Dactylorhiza*. PFA is the result of burning powered coal in a power station boiler and is a fine, grey-white powder whose particle size distribution is comparable to talcum powder. Its chemical properties vary with its age; when fresh it is strongly alkaline (pH>9) and with saline (<=30mS cm$^{-1}$; *c.* ½ salinity of sea water) leachates whose boron content renders them phytotoxic to many plants. Exposure to rainwater and weathering ameliorates these extreme conditions, leaving a material comparable to a calcareous sand within about five years.

## Introduction

Fresh PFA is notoriously difficult to revegetate, being alkaline, phytotoxic and largely devoid of fertility (especially nitrogen). Consequently, for the first few years after dumping PFA remains largely bare and its early plant colonisers are mainly salt-adopted annuals and ruderals (*Atriplex prostata*, *Puccinellia* spp, *Plantago coronopus*, *Spergularia marina*): *Sysimbrium altissimum* and *Chenopodium* spp (including scarce oak-leaved goosefoot *C. glaucum*) are also commonly encountered. Once salinity has declined to <1mS cm$^{-1}$ (3–5 years) a range of forbs grasses and legumes start to colonise; by 10 years the surface is usually fully vegetated, although the tops of mounds may remain bare for longer due to lack of water. It is during the grass/legume colonisation that Dactylorhiza orchids usually first appear, though they are easily overlooked at this stage due to their low numbers. Throughout England and Wales the same three species of orchids co-appear on PFA: *D. incarnata*, *D. praetermissa* and *D. fuschii*, along with numerous indeterminate hybrids. (At any one site it si common for one of these three to dominate but the balance may differ dramatically and unpredictably between apparently comparable sites.) The only other orchid the author has yet seen on PFA is *Ophrys apifera*, once on one site in the north-west. Once established, orchid numbers can explode dramatically, with thousands of slower spikes appearing in a few years, making for an unforgettable spectacle.

The succession proceeds onwards to scrub woodland, usually *Betula* and *Salix*, although *Hippophae rhamnoides* has a proven ability establish dense monocultures on young PFA. As the canopy closes flora diversity drops sharply (though this is not true for soil invertebrates; other groups still need research). Where open glades persist these often contain remnant orchid colonies, sometimes with the round leaved wintergreen *Pyrola rotundifolia*. The succession has strong analogies with the sand-dune/dune slack system, although sand dune grasses generally colonise poorly on fresh PFA.

A recurring pattern is for PFA to be deposited as an unwanted waste and abandoned for many years, during which time an extensive and spectacular orchid colony develops into a noted local feature. Realisation of the site's interest often coincides with its decline as a canopy closure casts an ever-denser shade, and even given severe scrub-removal, orchid colonies tend to fade away as the site ages; few sites over 40 years have orchid colonies of note.

The work focused on three PFA-orchid sites; two were experiments set up in 1991 to explore the promotion of orchid establishment and the third was a large public site in the Lee Valley. In the latter, where orchids had been a prominent local feature but were then declining.

**Site 1, Tilbury Power Station, Essex**: A specially prepared are of PFA 25 m × 25 m × 0.5 m deep was inoculated in March 1991 with soils scraped from existing PFA sites of various ages, including the Lee Valley site. *D. incarnate* appeared in small numbers in 1998, with an explosion in numbers, starting in 2002. The first few plants (1998–2001) were all in plots inoculated with PFA from the Lee Valley PFA sites but the second wave of orchids appeared in a variety of treatments but always near the centre of the site. The edges of the site had been invaded by thick rank vegetation, while the orchids (almost entirely) *D. incarnata* grew in the thin sparse turf remaining in the plot centre. This central region was lower in nitrate than elsewhere. A manual scraping exercise was introduced in winter 2002 to remove dense vegetation, intending to give the orchids fresh PFA and seed into. It was too early to assess whether this was successful, but the ruderals colonising the newly-scraped area were almost entirely different to those found on freshly deposited PFA.

**Site 2, Drax Power Station**: Here, six PFA-containing mounds were inoculated with PFA surface soil in 1991. Orchids (mainly *D. praetermissa*) first flowered in 1995, and exploded in 2000 (increasing steadily from 20 in 1999 to >700 in 2003). Where experimental treatments were imposed the mounds became a dense grass/scrub mix with few orchids, but the untreated areas of the mound bases remain almost bare, with potential for vast orchid numbers, albeit transiently.

The early stages of colonisation at Sites 1 and 2 are documented in Shaw (1996).

**Site 3, Northmet Pit, Chestnut, Lee Valley**: Here, an area of 50 m × 300 m × >2 m in depth of PFA was deposited in the 1960s, giving rise to spectacular displays of all three PFA orchid species, visible from public board walks (Shaw 994). Displays peaked in the 1980s and have gone into dramatic decline (from thousands to single figures in <10 years).

It is proposed that the only way to maintain this site as an orchid colony is a dramatic re-setting exercise (Shaw, 1998) in which woodland and topsoil are removed and fresh PFA is deposited in selected areas to re-start a primary succession. The ethics of moving orchid germplasm around the UK were briefly discussed.

# References

Shaw, P.J.S. (1994) Orchids woods and floating islands – the ecology of fly ash. *British Wildlife*, **5**, 149–157.

Shaw, P.J.A. (1996) Role of seedbank substrates in the revegation of fly ash and gypsum in the UK. *Restoration Ecology*, **4** (1), 61–69.

Shaw, P.J.A. (1998) Conservation management of industrial wastes. *Practical Ecology & Conservation*, **2**, 13–18.

# Book Reviews

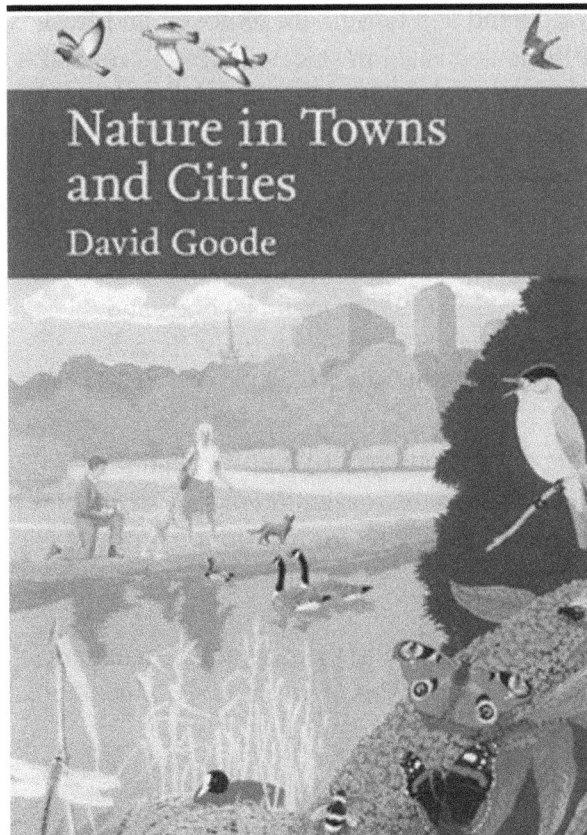

**Nature in Towns and Cities**, by David Goode
First published, 2014, by Harper Collins
Publishers, 77-85 Fulham Palace Road,
London W6 8JB ISBN 978-0-00-724329-9
(Hdb.) Price £55

The Collins New Naturalist series has a
deserved reputation as a leader of the genre
over many decades. The accolades come
from the wider, popular media as well as
the specialist natural history or academic
audiences. So for example, '*Taken individually
or as a whole they are one of the proudest
achievements of modern publishing*' The
*Sunday Times*; '*The series is an amazing
achievement*' The Times Literary Supplement;
and '*The books are glorious to own*' The
Independent.

The New Naturalist books tend to be excellent
natural history writing rather than academic
ecology, and that is not such as bad thing.
Furthermore, they seem to at least try to cover
all aspects of British natural history. However,
there are certain themes which the series has
sometimes struggled to get to grips with,
and urban ecology is perhaps one of these.

With David Goode's superb monograph the
New Naturalist label has been thoroughly
urbanized with a volume to sit alongside
the classic account *The Ecology of Urban
Habitats* written by Oliver Gilbert for Springer
(Chapman & Hall) back in 1989.

David Goode was, I suggest, possibly the
only person capable of writing such a volume
that spans the remarkable scope of the urban
and urbanized environments. David has a
long and distinguished career in ecology and
nature conservation since graduating from the
University of Hull in Geology. He went on to
take the then pioneering postgraduate Diploma
in Ecology and Conservation at University
College London and gained his doctorate in
bog ecology. He joined Nature Conservancy
in 1967 as the national specialist for peatland
ecology and conservation, and later became
Assistant Chief Scientist. However, in 1982,
he moved into the urban scene to become
Senior Ecologist at the Greater London
Council, then Director of the London Ecology
Unit, and latterly Head of Environment at The
Greater London Authority.

In these various roles, David was directly
involved in developing the theory and practice
of urban nature conservation. This activity
was as both a professional ecologist and an
enthusiastic naturalist with particular interest
in birds. He wrote *Wild in London* (1986),
jointly edited two other books and written
many popular articles and papers on urban
ecology. A Visiting Professor at University
College London since 1994, he was Honorary
Professor at East China Normal University
from 1996-2001 and is a member of the IUCN
Specialist Group on Cities and Protected
Areas. He is a Fellow, Past President and
Patron of the Chartered Institute of Ecology
and Environmental Management. In 1999,
he was awarded the international Heidelberg
Prize for Environmental Excellence.

Today in our increasingly urban society, we
are all familiar with towns and cities. Indeed,
most people live in and around urban centres.

Yet though most of us live in such areas, many are unaware of the rich ecology with which we share the city space. There is now no excuse for ignorance since with this extraordinarily wide-ranging study, David Goode has drawn on many years' research, meticulous observation, and experience to explain the diverse habitats and species of the urban landscape.

Many modern urban ecologists had their interest sparked by remarkable book by Richard Fitter published in 1945, London's Natural History. This was Volume 3 in the New Naturalist series and, though focused on the capital, was a substantial contribution to urban ecology. Since that time, there have been major developments in an expanding knowledge and understanding of urban ecology. In particular, we know much more about how animals and plants respond and adapt to urbanization. Alongside this growth in understanding has been a spectacular rise in urban nature conservation, and David presents a compelling account of how has come about.

This is a long-awaited addition to the New Naturalist library and a huge contribution to urban natural history writing. Goode applies his knowledge of urban wildlife demonstrate how nature is re-colonising the town, and urban ecological science is informing conservation. With a vast experience of practical planning and development issues, David shows how urban ecology can inform site management to good effect in the city. The core ecological principles can be applied to design and planning approaches to conserve encapsulated countryside or to create new woodlands, wetlands and grasslands. Such new approaches to the urban landscape have developed in many countries over the last thirty years and with globalisation and urbanisation on a massive scale today, the significance can only grow in the future.

Above all, the book is written with knowledge, passion and humour, and those working on urban trees may welcome this as a break from the confines of their own core disciplines. The volume is well illustrated with colour images of habitats and species and the book

will help the urban dweller or worker to see and appreciate urban nature. All-in-all, this is natural history writing at its best. So put your feet up and read.......!

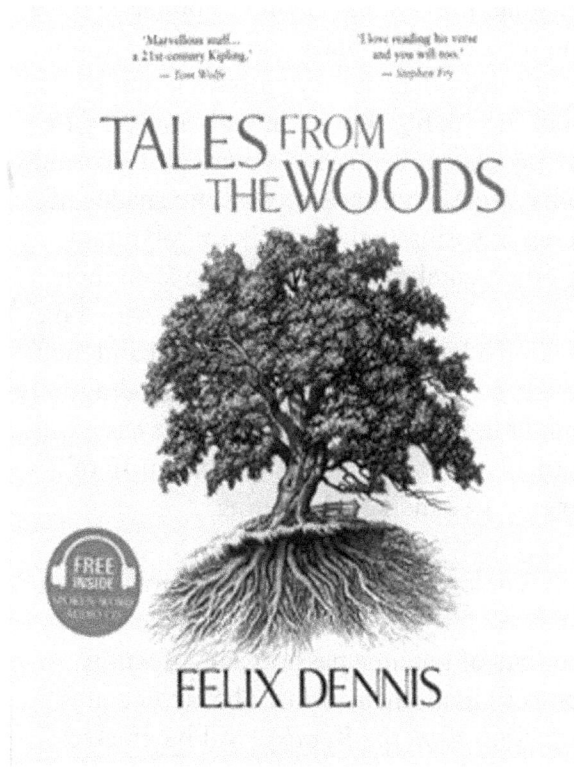

***Tales from the Woods***. By Felix Dennis. Published by Ebury Press, 20 Vauxhall Bridge Road, London SW1V 2SA, 2010. ISBN 9780091937676 (Hbk.) Price £9.99

Felix Dennis (b. 27 May 1947, d. 22 June 2014), was an English publisher, poet, spoken word performer and philanthropist. His company, Dennis Publishing, pioneered computer and hobbyist magazine publishing in the United Kingdom. He began his business and writing career in the hugely controversial magazine Oz, in the 1960s and early 1970s. Later, his company added lifestyle titles such as its flagship brand The Week, published in the UK and the United States. Dennis had a passion for trees, and by the early 2000s, for poetry. It is these twin obsessions that are celebrated in what can only be described as a 'lovely book'. This is a long way from the 1970s trial at which he and his co-writers were charged (unsuccessfully) with 'Conspiracy to deprave and corrupt the Morals of the Young of the Realm'. They were convicted and sentenced to imprisonment for two lesser offences, though Dennis got a more lenient sentence as he was considered by the judge, to be 'much less intelligent' and therefore less culpable of the offences.

Tales from the Woods is described as 'a thrilling, impassioned hymn of praise to trees and the English countryside from one of our best-loved poets. His eye misses nothing - from the quality of light in early morning to the joyous exuberance of young birds learning to fly; from a hornbeam's delicate shades of green in spring to the 'kamikaze' obduracy of ornamental maples refusing to shed their leaves in November gales. Here is 'Mrs. Fox' making a home for her cubs in an abandoned cellar; a spider attacking a sea-green dragonfly caught in its web; an apple tree that has learned to 'walk'; and the terrible curse of a veteran tree marked for felling.'

Bill Sanderson's beautiful illustrations adorn the book throughout, and this adds to the feelings of genuine passion and affection for trees in all their forms. The cover states that *Tales from the Woods*, 'will be greeted with delight by all lovers of trees and the countryside – and especially those who have already fallen under the spell of a poet described by one eminent critic and author as "a 21st-century Kipling".' It also comes with a cd of readings by Dennis of his own work. Over the years, he has acquired a remarkable following of readers and enthusiasts, no doubt aided by his early days in the rock and pop underworld. Comments on the book cover are from a stellar cast include: Christopher Rush: 'As with Pablo Neruda, he has that simple ability to throw language like a dress over experience and make the mundane marvelous.'; Mick Jagger: 'I enjoy his poetry immensely.'; Richard Fair: 'Total, utter joy – so real, so readable and so enjoyable.'; Stephen Fry: 'I love reading his verse and you will too'; Tracy Farnsworth: 'The most satisfying collection of poetry I have ever read.'; Shirley Conran, OBE: 'He invokes sorrow as fast as regret, pain as readily as passion, and love as tenderly as murderous rage.'; Jon Snow: 'A fantastic collection. Rich, sumptuous and beautifully threaded.'; and even Sir Paul McCartney: 'His poetry sings like a summer breeze through the fairground.' In spite of the somewhat gushing accolades

above, this is a book to acquire and at £9.99, it is a bargain. Turn to pages 72 to 73 and enjoy a wonderful illustration of a gnarled, old oak tree with a short poem *Veteran*:

What have you seen? Whom did you shade?
How many winters have battered your boughs?
How many seeds did you sow in this glade?
How many beetles and birds did you house?

Where is your crown? Shattered and flayed?
Fear no usurper to topple your throne!
Here you shall stand in your royal glade.
And the day of your falling shall be your own.

There follows a short observational text about how we interfere too much in veteran trees, and how health and safety rules. As Dennis says, 'poets are mad, bad and dangerous to know. Just like veteran trees!' The poem and the text should be read by anyone who manages old trees.

Tales from the Woods is a remarkable book and a testament to a great passion. Anyone interested in trees will find this a pleasure to pick up and browse during quiet moments by the fireside Felix Dennis was a publisher, poet and planter of trees. His five previous books of poetry made him one of Britain's best-loved poets and his poetry has been performed by The Royal Shakespeare Company on both sides of the Atlantic. His hugely popular poetry tours were the subject of a forthcoming documentary, *Full House*. He lived in England, the USA and on the Caribbean island of Mustique.

# IUER Editorial Board